The Chemistry and Technology of Printing Inks

Norman Underwood, Thomas V. Sullivan

The Chemistry and Technology of Printing Inks

BY

NORMAN UNDERWOOD

CHIEF OF THE INK-MAKING DIVISION, BUREAU OF ENGRAVING AND PRINTING, UNITED STATES TREASURY DEPARTMENT

AND

THOMAS V. SULLIVAN

ASSISTANT CHIEF OF THE INK-MAKING DIVISION, BUREAU OF ENGRAVING AND PRINTING, UNITED STATES TREASURY DEPARTMENT

Illustrated

NEW YORK

D. VAN NOSTRAND COMPANY

25 PARK PLACE

1915

PREFACE

The authors have endeavored in the preparation of this volume to prepare a concise work on the chemistry and methods of manufacture of one of the most important materials of the present day.

They have attempted to give in a brief and practical but yet scientifically correct manner the many facts concerning the raw materials and finished products used in this industry which they have collected during a number of years of laboratory work and manufacturing experience.

Obsolete methods and materials which have been found to have no value in the art on account of modern improvements or excessive cost have been omitted.

The attempt has been made to present the most recent methods of manufacture and a description of the materials which have been found useful in the art in a clear and concise manner. The authors have spent a great deal of time on the form and style of the book in the hope that it may prove valuable and serviceable to the many workers in this art.

WASHINGTON, D. C. — N. U.
Sept. 1, 1914. — T. V. S.

CONTENTS

INTRODUCTION

The manufacture of printing inks today, does not, as would appear at first glance, consist merely in the combining together of pigments, driers and vehicles in certain proportions.

Besides the mere combination of ingredients, the quality, suitability and characteristics of these materials must be considered in order to arrive at the combination that will be both chemically and physically adapted to give the best results.

In order to determine the quality, suitability and character of the ingredients that go into printing inks, it is necessary to have some knowledge of the constitution and methods of manufacture of these materials, and to be able by chemical analysis and physical tests to determine whether these materials have been properly made and whether they are by nature fitted for the work they will be called upon to do.

If the characteristics and properties of all the raw materials employed in the printing-ink industry are well understood and one has a comprehensive technical knowledge of the results to be obtained by the use of a certain ink, combinations can be worked out that will give simple formulas and ones that will give the very best working inks for any given purpose, with the added advantage that they will be easy to manipulate to meet any unexpected conditions that may arise in printing.

With these facts in mind, we have divided this book into three parts: namely, methods of testing raw materials; the manufacture and properties of pigments and varnishes; and the manufacture of inks.

We have not devoted any space to the chemical analysis of finished inks as there is nothing to be learned by such an analysis, except, that in some cases it may give one thoroughly conversant with inks some slight suggestion in regard to the duplication of an ink. A general chemical analysis will not, however, convey any idea of the essential composition of the ink. It will show the percentage of oil to pigment, whether the oil is linseed, rosin, or petroleum oil, and in a general way what the pigment is, but it shows practically nothing concerning the physical condition of the materials before manufacture or how they were combined; and unless the chemist has a practical knowledge of ink making, it does not show even this much.

For example, an analysis may show the presence of linseed oil, but it will not show the consistency of the oil used in making the ink nor whether two or three different consistencies were used. Neither will it show if the oil was combined as a gum varnish, in a drier, or whether it was burnt or boiled. It will not show whether the drier used was a paste drier, a japan drier, a drier merely added dry when the ink was made or incorporated into the oil at the time of burning or boiling, or whether it was introduced in the form of a metallic soap. It may show the presence of a mineral oil, but it will not show whether the oil is petrolatum, paraffine oil or kerosene. If aluminum is found, it will not show whether it was added as a base to the ink, or whether it was part of a lake pigment that was used.

The following analyses as reported by analysts not familiar with the manufacture of inks together with their actual composition will be of interest in this regard.

ANALYSIS OF AN ENGRAVING BLACK

Oil (partly boiled linseed)	30.31
* Carbon	38.90
* Silica and insoluble silicates	9.50
* Calcium Phosphate	9.04
* Calcium Sulphate	2.82
* Calcium Carbonate	1.83
Oxide Iron	.26
Berlin Blue	7.34

Items marked * represent 62.35% of common spent bone black.

Now this ink was actually composed of the following materials:

Bone Black	37 %
Vine Black	16 "
Prussian Blue	7 "
Paste Drier	10 "
Strong Burnt Plate oil	4 "
Medium Burnt Plate oil	26 "

In this ink the vine black and strong plate oil are essential for the proper color and working properties, yet their presence is not shown by the analysis and could not be deduced from it.

ANALYSIS OF A BLACK TYPOGRAPHICAL INK

Pigment	20 %
Oil	80 "

Oil

Rosin oil	60 "
Rosin	22 "
Linseed Varnish	18 "

Pigment

Ash and oil driers	2.5 "
Prussian Blue	3.9 "
Carbon, etc.	93.6 "

This ink actually contained only about 4 per cent of linseed oil, the basic vehicle being a varnish made from paraffine oil and rosin, the rest of the vehicle consisting of about 15 per cent second run rosin oil and about 20 per cent of a long varnish made from paraffine oil, asphaltum, wood tar and kerosene with a small amount of japan drier. An ink made from the formula as shown by the analysis would not make an ink that was suitable for the purpose this ink was intended, that is for a news ink run on a high-speed web press on news paper.

In our work on printing inks and ink-making materials, we have found that the simpler a formula is, the less trouble one will have with the ink, and in the event of trouble, it will be easier to identify the cause and remedy the defect. There has always been a tendency in the printing-ink business, as conducted in a practical way, without reference to the technical side of the industry, to have formulas for inks, particularly those for special purposes, as complicated as possible. This is due mainly to the fact that these inks were the result of long "rule of thumb" experiments, in which the different materials were added until the combination sought for was obtained, and partly to the fact that every ink maker and printer has a number of trade secrets which are considered to be of great value in making inks work properly. The value, however, of these secrets is in most cases doubtful.

In the chapter on the manufacture of printing inks, we have not given any working formulas, since the properties of materials from different sources are quite variable and the conditions under which inks are used are liable to be very different. It would therefore be impossible to give exact figures that would be of any service and we have only given general formulas for the composition of the usual classes of inks, and an outline of the defects liable to occur and the remedies that we have found effective.

The information in this book represents a great deal of research, both in the laboratory, and in the works, nothing being taken for granted, and no statement is made that has not actually been proved in practice. In order to avoid useless discussion and to make the work simpler, we have only mentioned those things which we have found to be the best of their line. There are a great many materials and methods used both by ink manufacturers and printers, to accomplish certain things which we have tried, but have not mentioned, either, because they did not accomplish the result claimed for them, or because we found something else would do so better or in a simpler way.

The pigments to be used for the manufacture of printing inks must be looked at from a different standpoint and judged by different standards than those to be used in making paints. While a great many pigments of the same chemical composition are common to both industries, the fact that one of them under certain conditions is acceptable as a material for paint, cannot be taken as an indication that it will be equally acceptable when ground into a printing ink. There are a great many pigments, that, while they are available for use as paints, cannot under any circumstances be used for ink and a number of very satisfactory ink-making pigments are not, for various reasons, adaptable for use in making paints.

The chief source of these differences can be easily explained by the ultimate use of the two products, as it can readily be seen, that the materials which are used to make a product to be applied with a brush, in a thick layer, and whose chief function is covering, can differ widely in properties from those that are to be used in a product which is put on in the thinnest possible layer and which has many other functions to perform besides that of covering surface.

While a pigment for use in paint must be judged, to a certain extent, by its color strength, its body, its fineness of grinding, its miscibility with other pigments and the usual vehicles, its durability and weathering properties and its fastness to light, a pigment for use in printing ink must be examined into for all of these properties in the minutest way; and it must also have certain other characteristics not necessary, and in some cases not desirable in paint pigments. For example, there are pigments that, when mixed with a large quantity of oil and thinned with turpentine, work well under a brush but show shortness and lack of viscosity, when made up to the consistency of ink, and which will not distribute at all on the press. Venetian red, an admirable paint pigment, is, on account of its hard, gritty nature, of practically no value as an ink-making pigment. A large number of aniline lakes, on account of their transparency and lack of covering power, are not used in paints, but for this very reason are desirable for use in certain kinds of printing inks.

In the following chapters we will take up these pigments and discuss them only from the standpoint of their availability for use in printing inks, making no mention of those pigments not suitable for this work.

Explanation of the Terms that are to be Used. — As many of the technical terms used to describe the properties of pigments and inks are frequently used in a confusing and indefinite way, the following explanation of the meaning of the terms hereafter to be employed, will help to eliminate any misunderstanding in their use.

Hue. — As the normal spectrum colors merge into each other, we have a condition where one color has a slight mixture of the other in it and this slight mixture gives to the predominating color what is called a hue. Thus if we have red slightly tinged with violet, this is called red of violet hue. On the other hand if we have a violet merging into red, that is, if a slight amount of red tinges the violet the result is a violet of red hue. Green in passing into blue gives us first a green of blue hue, and then a blue of green hue, as first the one, and then the other predominates.

In speaking of pigments this can be carried further, to mean that, when two pigments are mixed to produce a new color and the color of one pigment predominates in the mixture, the resultant color will have the hue of the predominating pigment. Thus when yellow predominates in a mixture of yellow and blue, we get a green of yellow hue and if the blue predominates, we get a green of blue hue.

Tint. — When a normal spectrum color or hue of that color is mixed with white, we get a gradation of that color lighter in appearance and this is called a tint of the original color. In speaking of pigments a tint means a pigment lightened by the admixture of white.

Shade. — When a normal spectrum color, or one of its hues, is mixed with a small amount of black or its complementary color, we get a gradation of that color darker in appearance than the original and this is called a shade of the color.

The term shade is frequently misused to convey the idea of hue. In the following pages the word shade will be used to convey the idea of a pigment darkened by the addition of black or another pigment complementary to it.

Top Hue and Under Hue. — When different thicknesses of a medium are used, there is in some cases a different amount of color absorption for the varying thicknesses of the medium and this causes a variation in the color or the hue of the medium. Thus a dilute or thin solution of magenta shows a pink of blue hue, while a solution of very great depth or strength is a deep red. Cobalt glass when very thin, shows a blue color, while as the thickness increases, it becomes purple and finally if a sufficiently great thickness be employed the color is a deep red.

The cause of this action is that there is an unequal coefficient of absorption between the different densities. At varying densities the absorption curve takes in or excludes different parts of the spectrum and the result is that the apparent color is of different hues, depending on the proportionate increase or decrease of the absorption.

When the medium is an opaque pigment and is spread on paper in a solid band so as to completely shut out the white of the paper or if the pigment is viewed in a mass, it shows the absorption that a dense solution would show; but when diluted with a white pigment or printed thinly on paper we get the same condition of unequal absorption as occurs in dilute or thin solutions, and a hue of the original color is the result, the white pigment or the white light of the paper acting as a dilutent. On account of this we often get a red of yellow hue in the mass while on rubbing or printing it out thinly or diluting with white we get a red of blue hue. From this explanation it can be readily seen that we can have a number of these variations between the material in mass and the same material diluted, or printed out thinly. The hue of the color when taken in the mass or undiluted we will call the top hue while the diluted or thinly spread out hue we will call the under hue of a color.

Color Strength. — By the term color strength of a pigment is meant the actual amount of color it contains and is measured by its tinting power when mixed with white. This of course is a measure of the relative power a pigment has to impart its hue or color to another pigment. The color strength of a pigment varies inversely with its crystalline character, the more amorphous a pigment is the greater will be its color strength.

Abrasive Quality. — If a pigment contains any hard material it is apt to exert an abrasive action on plates, forms and cuts and to wear down the fine lines so that sharpness and definition will be lacking. This not only destroys the effectiveness of the work but also shortens the life of the plate or cut.

Fineness. — This word is used in its ordinarily accepted meaning. In this connection it is well to note that taking the term abrasive quality in the sense shown above, a material can be quite fine and at the same time can be abrasive. A great degree of fineness is necessary in all pigments to be used in ink making not only on account of the fine lines and delicate tone effects which would not be brought out unless the pigment was ground very finely but also because a coarse pigment has a tendency to collect into balls and to give a granular effect to the ink. Precipitated colors frequently appear coarse but when rubbed under the finger this apparent coarseness disappears.

Oil Absorption. — On account of differences in the physical structure of pigments the amount of oil necessary to make a thin paste varies, some pigments taking more than others; the measure of this quality is called the oil absorption of a pigment.

Livering. — When an ink, on standing, thickens to a spongy, rubber-like mass it is said to liver. This is due to a chemical action between the pigment and the vehicle, such as the rapid oxidation of the oil, or the formation of a soap. When a pigment shows a tendency to liver it cannot be regarded as a good ink-making pigment.

Shortness. — If a pigment when mixed with a large quantity of oil still remains stiff or cannot be drawn out into a string between the fingers but breaks, it is said to be short. While there are some classes of work that require an ink of a certain degree of shortness, as a general rule, pigments that show this property are not suited for making inks.

Flow and Length. — Flow is the property of a pigment to combine with a good body the abilty to run and feed well on the press. An ink that flows well must also have the property of being drawn out into a string between the fingers and this is called length. Thus each of these terms suggests or includes the other; they are both sometimes spoken of under the name of viscosity. For all classes of work these properties are very essential and a pigment that does not show them to some extent should be avoided.

Tack and Softness. — Tack is that property of cohesion between particles of ink that can best be described as the pulling power of the ink against another surface. When there is very little cohesion and the property of tack is almost absent we have what we shall designate as softness.

Body Color. — Body color is the color of the dry pigment before it is mixed with a vehicle. The addition of a vehicle frequently makes a decided difference in the color of a pigment.

Transparency. — Transparency will be used in the general sense of the word, namely a pigment or ink that allows color or light from another source to pass through it.

Opacity. — Opacity is the property of absolutely stopping the transmission of light or color from another source. A transparent color can be rendered more or less opaque by the addition of a base, while many opaque pigments can be made to some extent transparent if they are laid on or printed in a thin film and some few can be made to appear transparent to a slight degree by the addition of other materials. The opacity or covering power varies as does the color strength with the crystalline character of the pigment.

Body. — The body of a pigment is the measure of its density. This term is variously used, generally to convey the idea of a property not readily explainable. Thus it is sometimes spoken of to convey the idea of covering power, and this is due to the fact that the denser an ink is the more covering power it will have. As regards its use in this volume it is the measure of the consistency and density of the ink. Thus a stiff ink that stands is said to have a good body while an ink that is soft and runny is said to lack body.

Incompatibility. — Incompatibility means that for some reason, physical or chemical, a pigment cannot be used with another pigment or with certain vehicles or under certain circumstances. The fact that we cannot use lead colors with those containing sulphur is a well-known example of this. Another illustration is the fact that colors affected by alkalies should not be used to print labels for soap or lye cans.

Bleeding. — Certain pigments when mixed with water or oil or any of the various printing-ink vehicles are partially soluble and this solubility is called bleeding. Pigments that bleed are not of much value for making printing inks as they are apt to strike through the paper or color the edge of the work. A pigment that is to be used in dry work need not be entirely insoluble in water but as a general rule a pigment that bleeds has not been properly made.

Fastness to Light. — This is used in the ordinary sense of the word, that is, to mean the degree of resistance the color has to the changing action of ordinary light.

Atmospheric Influences. — Closely allied to fading and sometimes attributed to it, are other changes both physical and chemical caused by the influence of the atmosphere.

These may be produced in various ways, by oxidation, reduction, the solvent action of gases and by chemical combinations between the pigment and acid radicals present in the air or on account of changes in the physical form of the color, due to dryness, moisture or extremes of temperature. The action of sulphur gases on lead pigments and the greenish hue that prussian blue assumes from exposure to oxidizing influences are common examples.

PART ONE

TESTING OF MATERIALS

GENERAL LABORATORY APPARATUS, CHEMICAL AND PHYSICAL TESTS

SECTION ONE. LABORATORY APPARATUS

The following methods of analysis and tests of ink-making materials have been compiled from various sources too numerous and varied for us to attempt to give any credit. All of them have been practically tried by the authors and found to be satisfactory.

As the results obtained in testing depend entirely on the accuracy and uniformity with which the work is done and as this accuracy and uniformity depend in a great measure on the apparatus employed, a brief description of the standard apparatus employed by the authors in testing inks and ink-making materials will be given. Some of this apparatus is in general use for testing pigments, inks and paints, some has been adapted from other sources and some is original with us.

Muller and Slab. — The slab used for rubbing colors should be of smooth polished marble, perfectly level, set in a heavy framed table that must be perfectly rigid. The muller should be made of some close-grained, hard material such as lithographic stone or a smooth, fine-grained marble. It should be conical in shape to give a good purchase to the hands and to allow of its being shoved with the maximum of downward pressure. The one in use in our laboratory is 8 inches high, 3 inches in diameter at the bottom and weighs about 4 pounds. The place gripped by the hands is slightly roughened to give a good purchase and the bottom is highly polished. The surface of the slab must be absolutely true and smooth so that the pressure will be uniform and to avoid having the particles of color gather where they will not get sufficient rubbing. A cut of the slab and muller, showing the color rubbed out, is shown in Figure 1.

FIGURE 1. — MULLER AND SLAB, SHOWING METHOD OF RUBBING OUT COLORS.

Electric Muffle Furnace. — An electric muffle furnace will be found invaluable, as with it almost any range of temperature can be obtained and it is far superior to the bunsen burner for ashing materials and burning precipitates. It is easily regulated and will reproduce at any time a definite condition of temperature used previously, so that work can be duplicated at different times under exactly similar conditions.

Electric Combustion Furnace. — An electric combustion furnace will also be found to be far more satisfactory than the ordinary gas combustion furnace.

Constant Temperature Oven. — The most satisfactory constant temperature oven is an electric one. The one we use has an easily controlled range from 75° to 150° C.

FIGURE 2. — LABORATORY MIXER.

There are also electrical hot plates and heating apparatus of almost every description which are more satisfactory than the gas burning ones, and by the use of these electrical devices in a laboratory the danger of fire is almost entirely eliminated.

Mixing Machine. — For mixing small experimental batches of ink we use an electrically driven mixer, the picture of which is shown in Figure 2. This mixer takes an iron mixing pan 9½ inches in diameter and 4 inches deep, which will hold about 2 pounds of material. The dry color and oil are weighed directly into the tared pan and it is then set on the mixer.

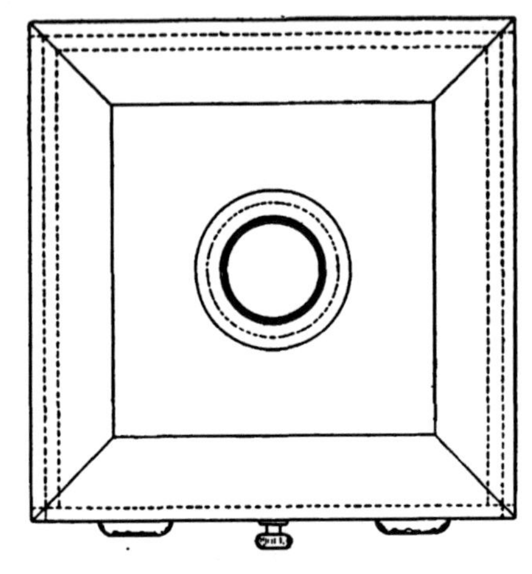

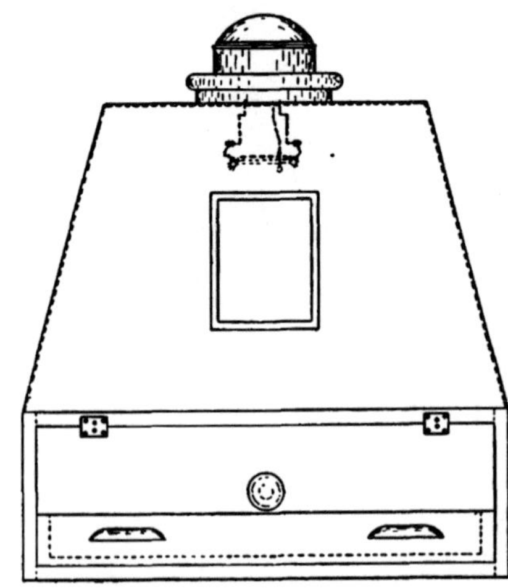

FIGURE 3.— ULTRA-VIOLET LIGHT AND CASE FOR TESTING THE EFFECT OF LIGHT ON COLORS.

Grinding Mill. — For grinding experimental batches of ink under the same conditions as will occur in ordinary work (and this is an important factor, as all laboratory work of this sort should be done as closely as possible under conditions that can be duplicated on a practical scale), we have a small mill made by the Kent Machine Works which is in every way a duplicate of their large three-roll ink mill.

Ultra-Violet Light. — For testing the fastness to light of various pigments and dyes we use an ultra-violet light. This light is mounted in a sheet-iron hood 3 feet by 3 feet at the base, sloping up in the form of a pyramid enclosing the lamp as shown in Figure 3. The bottom of the light is set 21 inches above the bottom of the hood, which is fitted with a hinged door and drawer for convenience in placing the samples. As the rays given off by this light are very injurious to the eyes, a look-in-hole is provided, high enough up so that the light itself can be seen, and is covered with a Hallauer No. 64 glass for cutting off the ultra-violet rays. This light develops about the same intensity as strong sunlight but has the advantage of being always uniform and can be used in any sort of weather.

Balances. — An accurate analytical balance is, of course a necessity. For rougher yet somewhat accurate work where small amounts are to be weighed a small "prescription" balance will be found useful as will also a balance of the "Robervahl" type with a capacity of from 5 to 10 pounds. For ordinary heavy work a small platform scale that will weigh down to ounces and up to 100 pounds is necessary.

Microscope. — A good microscope with a photographic attachment is also a great adjunct in work on pigments, especially in determining their physical characteristics and the bases on which aniline lakes are precipitated. This branch of work is only beginning to be recognized and given its proper place in the color testing laboratory.

Filtering Apparatus. — For filtering large experimental batches of color a small filter press may be used but this will not be found any more satisfactory than the more simple rectangular frame with pins from which a filter cloth is suspended.

Oil Tanks and Cups. — Several small copper tanks with covers and wide opening spigots should be used to hold oils and varnishes for use in the laboratory and we have found a few cups with tops that lift back and a lip fashioned after the ordinary molasses pitcher are very useful when weighing out oils, varnishes and other liquids.

Glass Plates. — For testing and comparing color strength and top hue a thin colorless glass plate similar to those used to make photographic slides should be provided.

Thermometers. — The laboratory should be provided with sufficient thermometers of various ranges and an accurate pyrometer will be found of great use in determining high temperatures.

Hydrometers. — A set of 6 hydrometers divided as follows will be found to give a very useful range; Specific Gravity, .700 to .800, .800 to .900, .900 to 1.000, 1.000 to 1.200, 1.200 to 1.400 and 1.400 to 1.600. These instruments should be graduated to read .5 of a degree in the fourth place. A set giving Beaume readings for getting correct strength solutions for precipitations and the like is also necessary.

Viscosimeters. — The viscosimeter employed by us for the determination of viscosity in raw oils is the regular type of Engler viscosimeter, standardized against water. Figure 4 shows the one used for testing boiled or burnt oils and varnishes. It is a copper cylinder with a steam jacket so that the oil and jacket can be kept at a constant temperature of 100° C. Its inside dimensions are 12 inches by 4¼ inches, and it has marks so that the same amount of oil or varnish can be used each time. The amount of oil used in the one in use in our laboratory is

1200 c.c., and of this amount 1000 c.c. are run out in the test. The bottom of the cylinder is pitched slightly to make the flow even and the orifice out of which the oil flows is a stop cock with a $\frac{1}{8}$ inch opening fitted with a long handle so that it operates easily and quickly. The jacket is fitted with a thermometer as is also the top of the cylinder. The thermometer in the top extends well into the oil about in the center of the viscosimeter; the one in the jacket shows the temperature of the jacket. The apparatus should be leveled when set and should be fastened securely.

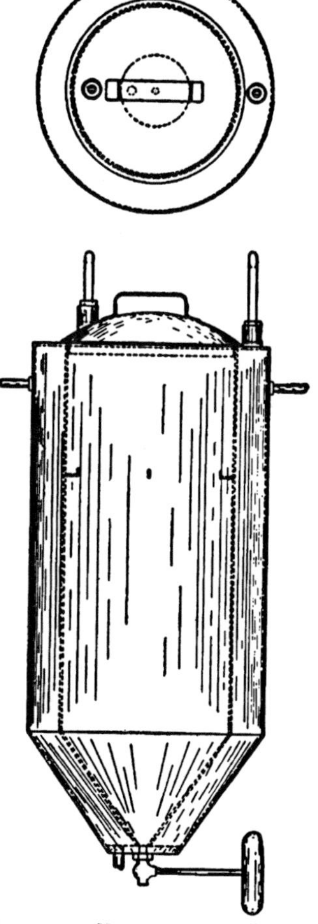

FIGURE 4. — VISCOSIMETER FOR TESTING HEAVY BODIED OILS AND VARNISHES.

We find this piece of apparatus invaluable in determining the consistency of oils and varnishes of heavy viscosity. The results are only relative, however, as the instrument is standardized against oils and varnishes known to be satisfactory.

Steam Jacketed Kettle. — A steam jacketed kettle cf the regular type holding about five or ten gallons will be found indispensable.

Crocks and Miscellaneous. — Several stoneware crocks from five to twenty gallons capacity should be provided for precipitating colors.

A motor-driven stirring machine, a number of heavy glass rods, smooth metal bars and wooden paddles, and several low wooden trestles should also find a place in the laboratory equipment.

SECTION TWO. METHODS OF ANALYSIS

Linseed Oil

Specific Gravity. — Specific gravity may be determined with a pyknometer or hydrometer at 15.5° C. For most control work an accurate hydrometer, reading to one-half a degree in the fourth place will be found as satisfactory as a pyknometer, providing the oil is uniformly cooled, the hydrometer immersed without touching the sides of the jar and the reading carefully taken. There should be no air bubbles in the liquid at the time of reading.

The specific gravity of a raw oil should be above .931, but, inasmuch as ageing, blowing and other permissible treatments that do not bring the oil into the class of a boiled oil, increase the gravity, it is useless to give a high limit for specific gravity. Boiled or burnt oils, or oils that have had driers added to them show a still higher gravity.

Viscosity. — The viscosity should be taken on an Engler viscosimeter at 20° C., water being taken as the standard. It is essential in order to get concordant results that both the water in the jacket and the oil remain at the proper temperature throughout the whole operation.

Flash. — The flash should be taken in a Cleveland open fire tester or some similar instrument in the following manner: The cup is put in the center of a ring of asbestos, extending all around it about 2 inches, and set on a tripod over a bunsen burner with the flame so regulated that the temperature of the oil will rise about 9° C. per minute. As a test flame, use an ordinary blow pipe attached to a rubber tube. Begin testing when the temperature of the oil reaches 250° C. and test for every rise of three degrees. When applying the test flame move it slowly across the oil in front of the thermometer a little above the surface of the oil. The flash is the lowest temperature at which the vapors above the oil flash and go out. Care should be taken that the oil is not foaming when the flame is applied, as the bubbles will burst and give an incipient flash which will, of course, be low.

Fire Point. — After determining the flash point, continue heating the oil until the vapors when ignited continue to burn. The temperature at which this occurs is called the fire point.

Drying on Lead Monoxide. — This test is similar to Livache's drying test, only litharge is used instead of precipitated lead. Weigh out about 5 grams of litharge into a tared watch glass, spreading the litharge on the bottom of the glass; distribute as evenly as possible over the litharge from .2 to .5 grams of oil. Take the exact weight of the oil and the exact weight of the watch glass, oil, and litharge, expose to the air and light for from 76 to 96 hours, weigh again and the gain is calculated in percentage from the original weight of the oil used. A chemically pure litharge must be used in this determination to get uniform results.

Acid Value. — Ten grams of oil are weighed into a 200 c.c. Erlenmeyer flask and 50 c.c. of neutral alcohol added, heat on a steam bath for one-half hour, cool and titrate with one-tenth normal sodium hydrate, using phenolphthalein as an indicator; calculate the acid value as milligrams of potassium hydrate per gram of oil. The acid value varies as does the gravity with age and blowing or treatment but should be less than 8.

Saponification. — Weigh 2 to 3 grams of oil into a 200 c.c. Erlenmeyer flask: add 30 c.c. one-half normal alcoholic potash, connect with a reflux condenser, heat on a steam bath for 1 hour; titrate with one-half normal sulphuric acid using phenolphthalein as an indicator. Always run two blanks with the alcoholic potash. From the difference between the cubic centimeters of acid required by the blanks and the sample calculate the saponification number in terms of milligrams of potassium hydrate to one gram of oil. The saponification number of linseed oil should be about 190.

Unsaponifiable Matter. — Saponify 5 grams of oil with about 200 c.c. one-half normal alcoholic potash for an hour on a steam bath, using a reflux condenser and evaporate the alcohol; take up in water and transfer to a separatory funnel, cool, shake out with two portions of 50 c.c. each of petroleum ether; wash the petroleum ether twice with water, evaporate the petroleum ether and weigh the unsaponifiable matter, which in raw linseed oil should not be over 1.5 per cent.

Iodine Number. — Weigh out from .2 to .25 gram of oil, transfer to a 350 c.c. bottle with a well ground glass stopper: dissolve the oil in 10 c.c. of chloroform and add 30 c.c. of Hanus solution: let it stand with occasional shaking for one hour: add 20 c.c. of a 10 per cent solution of potassium iodide and 150 c.c. of water and titrate with standard sodium thiosulphate, using starch as an indicator. Blanks must be run each time and from the difference between the amounts of sodium thiosulphate required by the blank and the sample calculate the iodine number (centigrams of iodine absorbed by one gram of oil).

The Hanus solution is made by dissolving 13.2 grams of iodine in 1000 c.c. of glacial acetic acid and adding 3 c.c. of bromine.

Foots and Turbidity. — Let one liter of oil stand in a clear glass bottle for eight days and note the amount of sediment formed. In very cold weather the oil will sometimes show a turbidity due to the freezing out of fats of very high melting point. It is therefore well in cold weather to heat a portion of the oil to about 100° C. and allow it to cool. If the turbidity does not return the oil should not be reported as turbid.

Ash. — Burn about 20 grams in a porcelain dish and weigh the ash. A pure raw linseed oil should contain only a trace of ash. In a boiled oil the ash will show the metallic driers. Make a qualitative test for these.

Break. — Heat 50 c.c. of oil in a small beaker up to 300° C. rapidly. If a jelly-like mass separates out, the oil has broken. This determination is influenced by the rate of heating to such an extent that it is difficult, unless the heating is done very rapidly, to get the same oil to act twice in the same way.

Drying with Manganese Borate. — One gram of oil and .2 gram of manganese borate are spread on a glass plate very thin and the time of drying noted.

Rosin. — (Liebermann-Storch Test.) To 20 grams of oil, add 50 c.c. of alcohol, heat on the steam bath for about fifteen minutes, cool, decant the alcohol, evaporate to dryness, add 5 c.c. of acetic anhydride, warm, cool and add a drop of sulphuric acid, 1.53 specific gravity. Rosin or rosin oil will give a fugitive violet color. The conclusiveness of this test is open to doubt.

Determination of Calcium Carbonate in Paris White

Weigh .5 gram of the sample into an Erlenmeyer flask, add 60 c.c. of fifth normal hydrochloric acid, close the flask with a reflux condenser about 2 feet long. Boil until steam comes from the tube; remove the flask and wash the tube into it, add a few drops of phenolphthalein and run in one-fifth normal alkali until the color just turns red. Calculate the calcium carbonate from the number of c.c. of alkali required. Lime may also be determined gravimetrically by precipitation in the following manner: .5 gram of the sample is dissolved in hydrochloric acid, the insoluble matter is filtered off, the filtrate is made alkaline with ammonia in slight excess and boiled. If iron and aluminum precipitate, the solution is filtered and washed with hot water, the filtrate to which more ammonia has been added is then boiled and a boiling solution of ammonium oxalate is added and the solution allowed to boil for about fifteen minutes, allowed to stand until settled and filtered; wash with hot water; burn off filter paper, blast for fifteen minutes and weigh as calcium oxide. By burning and weighing the insoluble residue the amount can be determined as can also the iron and aluminum oxides by burning and weighing the precipitate from the addition of ammonia. If it is desired to separate the iron from the aluminum this can be done by dissolving the mixed oxides in sulphuric acid, reducing with zinc and titrating with permanganate.

Magnesia may be determined in the filtrate from the lime as follows: evaporate in an acid solution to about 250 c.c. add about 25 c.c. of a solution of microcosmic salts, cool, add ammonia drop by drop with constant stirring until slightly ammoniacal; add about one-half the volume of ammonia and let stand over night; filter and wash with water containing ammonia and ammonium nitrate. Ignite and weigh as magnesium pyro-phosphate.

BARYTES

A microscopic examination of barytes should be made to determine the evenness of grinding, the size and angularity of the particles and whether it is amorphous or crystalline.

The only determination generally necessary for barytes is the amount of insoluble in hydrochloric acid. This insoluble can generally be considered barium sulphate but to make sure that there is no silica present, it is always well to evaporate with sulphuric and hydrofluoric acids.

Natural barytes is sometimes contaminated with calcium fluoride. Test qualitativly for fluorides by moistening with water in a platinum dish, adding sulphuric acid and gently heating. The top of the dish should be covered with a glass coated with paraffine with several marks scratched through to the glass. If fluorides are present the glass not covered by the paraffine will be etched. If there is any amount of soluble salts, iron, aluminum, lime, etc., they can be determined in the filtrate from the insoluble in the way outlined under paris white. If lime is present determine the soluble sulphate in 1 gram dissolved in hydrochloric acid and filtered. To the boiling filtrate add barium chloride, let stand, filter, ignite and weigh; calculate to calcium sulphate. If there is any lime left over and carbonates are present, calculate the remainder to calcium carbonate, unless the presence of fluorides is detected.

Lithopone

Barium Sulphate and Insoluble. — Dissolve 1 gram in hydrochloric acid, adding a small amount of potassium chlorate, evaporate to about one-half, dilute with water, add 5 c.c. dilute sulphuric acid, boil, allow to settle, filter, ignite and weigh. This is the barium sulphate and insoluble.

Total Zinc Oxide. — Heat the filtrate to boiling, add sodium carbonate drop by drop until all the zinc is precipitated as carbonate, filter on a Gooch crucible, warm, ignite and weigh as zinc oxide.

Zinc Sulphide. — Digest 1 gram in 100 c.c. of 1 per cent acetic acid at room temperature for one-half hour, filter and wash, determine zinc in filtrate by sodium carbonate. The difference between the total zinc oxide and the zinc oxide soluble in acetic acid, multiplied by 1.1973 gives the zinc present as sulphide.

Barium Carbonate. — Digest 2 grams with boiling dilute hydrochloric acid, dilute with hot water, filter, and determine barium in the filtrate by precipitation with sulphuric acid. Figure as barium carbonate.

White Lead

Insoluble and Total Lead. — Weigh one gram of the sample, moisten with water, dissolve in acetic acid, filter, wash, ignite and weigh the insoluble. To filtrate from insoluble matter, add 25 c.c. sulphuric acid (1: 1), evaporate and heat until acetic acid is driven off; cool, dilute to 200 c.c. with water, let stand for two hours, filter on a Gooch crucible, wash with 1 per cent sulphuric acid, ignite at low heat and weigh as lead sulphate.

Carbon Dioxide and Water. — In determining carbon dioxide heat 1 gram in a combustion furnace, catching the water in a calcium chloride tube and the carbon dioxide in a potash bulb; by weighing the residue left in the boat the exact composition can be calculated.

Calculate the carbon dioxide to lead carbonate, subtract the lead oxide necessary to form the carbonate from the total lead oxide present and calculate the remaining lead oxide to lead hydroxide. Calculate the water necessary for the lead hydroxide and subtract it from the total water and the remaining water can be reported as moisture.

In determining carbon dioxide the method in most common use now is to pass the evolved gas through a solution of barium hydroxide, filter as rapidly as possible with the greatest exclusion of air liable to contain carbon dioxide, wash in boiling water, dissolve in hydrochloric acid and determine as barium sulphate. Calculate the amount of carbon dioxide from the barium sulphate.

Calcium Sulphate

Determine sulphate by precipitation with barium chloride as described under barytes.

In another portion determine insoluble and in the filtrate from the insoluble determine lime by precipitation with ammonium oxalate.

Zinc Oxide

Determine zinc, after solution, as carbonate, as described under lithopone.

Chromium Colors

Chrome Yellow, Chrome Orange, Chrome Red

A microscopic examination will show the character of the pigment, whether it is crystalline or amorphous.

Dissolve .5 gram in hydrochloric acid, filter off insoluble, dilute the filtrate to 400 c.c., almost neutralize with ammonia, pass in a rapid stream of hydrogen sulphide until all the lead is precipitated as lead sulphide, filter, wash with water containing some hydrogen sulphide, dissolve the lead in dilute nitric acid, add an excess of sulphuric acid, heat to fuming and determine the lead as sulphate in the usual way.

Chromium. — Evaporate the filtrate from the lead sulphide to small bulk, add ammonia to excess, boil until the pink color disappears, filter, wash with hot water, ignite and weigh as chromium oxide.

Sulphate. — Weigh one gram, dissolve in hydrochloric acid, dilute considerably and determine the sulphate by precipitation with barium in the usual way.

Chrome Green

A microscopic examination will show the character of the particles and also if it is a dry mixture or made by simultaneous precipitation. A badly made green will show yellow and blue as well as green, while a well made green will show green and some blue but no yellow.

The analysis of the green is carried on in the same way as for chrome yellow except that the sample is first gently ignited in a porcelain crucible to decompose the blue.

The precipitate with ammonium hydroxide also contains iron besides the chromium, and it is necessary to dissolve this precipitate with hydrochloric acid on the filter and make it up to a definite volume. One part of this is taken and sodium peroxide added in sufficient amount to render the solution alkaline and to oxidize to chromate, boil until all of the hydrogen peroxide is driven off, cool, make acid with sulphuric acid, add a measured excess of standard ferrous sulphate and titrate the excess of iron with standard potassium bichromate.

Determine iron in another portion by reduction and titration with potassium permanganate.

VERMILLION

About the most satisfactory way to determine the purity of vermillion is to ash one gram of the pigment. Not more than .5 per cent should remain.

Vermillion should also be tested for dye toners by being shaken with alcohol, chloroform and hot water and filtered. If the filtrate shows color it is an indication of dye.

Iron Oxide and Earth Pigments

It is seldom necessary to analyze these pigments but when it is the iron and insoluble can be determined in the usual way.

Manganese. — For those pigments that contain manganese, dissolve .2 gram in nitric acid with heat and add 75 c.c. of strong nitric acid, boil and add 5 grams potassium chlorate, heat to boiling, add 50 c.c. more strong nitric acid and a little more potassium chlorate, boil until fumes cease to come off, cool, filter on asbestos and wash with strong nitric acid, suck dry and wash out the remaining nitric acid with water, transfer precipitate and asbestos to a beaker, add a measured excess of standard ferrous sulphate in dilute sulphuric acid, stir until all the manganese dioxide is dissolved and titrate the remaining ferrous sulphate with potassium permanganate. A ferrous solution of the proper strength is made by dissolving 10 grams of crystallized ferrous sulphate in 900 c.c. of water and 100 c.c. of sulphuric acid.

Ferrocyanide Blues

Prussian Blue, Bronze Blue, Chinese Blue

Insoluble. — Ignite one gram in a porcelain crucible at a low temperature, add 25 c.c. strong hydrochloric acid, boil until all the iron is decomposed, add water, bring to a boil again and filter, wash with hot water, ignite and weigh the insoluble.

Iron.— Make up filtrate to volume and in an aliquot part determine iron by reduction with zinc and titration with potassium permanganate.

Aluminum.— In another portion the iron and aluminum oxides may be precipitated with ammonia and the aluminum found by difference from the total oxides and the amount of iron found by titration.

Dyes. — It is well also to test the blues for dye toners by shaking them up in the various solvents. This should be very carefully done as it is sometimes very difficult to filter out the blue due to the formation of a colloidal solution. This is frequently taken to be a dye but in reality is the blue itself.

Ultramarine Blue

An analysis of ultramarine blue is of very little value. The principal determinations are, however, aluminum, silica, soda, total sulphur, sulphur as sulphate and sulphur as sulphide.

Silica. — Silica is determined by solution, evaporation to dryness, resolution and filtration, the residue being ignited and weighed.

Aluminum. — Aluminum is determined in the filtrate from the silica by precipitation with ammonium hydroxide.

Soda. — Soda is determined in the filtrate from the aluminum by adding sulphuric acid, evaporating to dryness and weighing as sodium sulphate.

Total Sulphur. — Mix 1 gram of pigment with 4 grams of sodium carbonate and 4 grams of sodium peroxide in a nickel crucible, cover with about a gram of sodium carbonate, fuse, using an asbestos shield to prevent the sulphur from being taken up from the gas; dissolve the fused mass in water, make acid with hydrochloric acid, precipitate with barium chloride and weigh as barium sulphate.

Sulphur as Sulphate. — Dissolve one gram in hydrochloric acid, boil till hydrogen sulphide is driven off, add barium chloride and weigh as barium sulphate. Subtract sulphur as sulphate from total sulphur and remainder is sulphur as sulphide.

Orange Mineral

Dissolve .25 gram of sample in 50 c.c. of a solution containing 1000 c.c. water, 126 c.c. nitric acid and 600 c.c. one-fifth normal oxalic acid. Then add 25 c.c. of 25 per cent sulphuric acid and titrate with permanganate; run a blank and calculate the difference to lead dioxide.

Blacks

Total Ash. — Weigh one gram into a platinum dish and burn off carbon at a very low temperature.

Insoluble Ash. — Transfer total ash to beaker, moisten with water, add 25 c.c. strong hydrochloric acid, boil for fifteen minutes, dilute with water, bring to boil again, filter off insoluble, wash with hot water and weigh after ignition.

Analysis of Soluble Ash

Phosphates. — Make up filtrate from insoluble matter to 500 c.c., if phosphates are present, draw off an aliquot portion, make alkaline with ammonia and acid with nitric acid, heat the solution on a water bath to 60° C. add 50 c.c. molybdate solution, heating at this temperature for one hour, filter and wash with a wash solution containing nitric acid, ammonium nitrate and a small amount of molybdate solution about three times. Dissolve the precipitate on the filter with ammonia, allowing the solution to run into the original beaker. Precipitate with hydrochloric acid and add ammonia in slight excess to redissolve the precipitate, cool, add 10 c.c. magnesia mixture and about one-half the volume of ammonia, stir, allow to stand over night, filter, wash with solution containing one part ammonia and three parts water, ignite, weigh and calculate to phosphoric anhydride.

Lime. — To another portion of the original solution add ammonia to alkalinity and make acid with acetic acid, filter, wash with hot water, boil the filtrate and add boiling ammonium oxalate, boil for a few minutes longer, allow to settle, filter, wash with hot water and weigh, after blasting, as calcium oxide.

Sulphates. — Determine sulphates in another portion with barium chloride.

Iron. — Take another portion of the original solution and add 15 c.c. of dilute sulphuric acid, evaporate to expel hydrochloric acid, take up with water, reduce with zinc and titrate with permanganate. When the black does not contain phosphates determine lime, iron and sulphates in the usual way.

Blacks are occasionally toned with prussian blue, and to determine this boil the sample with a 4 per cent solution of sodium hydroxide, filter, make acid with hydrochloric acid and add a solution containing a mixture of ferrous and ferric chloride or sulphate. A blue precipitate indicates the presence of prussian blue.

Organic Lakes

The only available schemes for identifying the aniline dyes used in lake pigments are much too complicated to be gone into in this volume. However the book entitled "Tests for Coal Tar Colors in Aniline Lakes" by George

Zerr and Dr. C. Mayer will be found to be a great aid in the examination of lake pigments.

SECTION THREE. PHYSICAL TESTS OF PIGMENTS

Color Strength. — .2 gram of the standard pigment and one gram of zinc white are weighed out accurately on an analytical balance and a like quantity of the pigment to be tested and zinc white is also weighed. Transfer these materials to a marble slab and add drop by drop sufficient oil to the standard to make it into a stiff paste. Add the same number of drops to the sample to be tested, mix both piles separately and thoroughly with a palette knife and rub each one 50 times with the muller described under laboratory apparatus. Each rub-out is then gathered up with a palette knife and rubbed again 25 times. After this final rubbing each sample is gathered together on the slab and a small amount of each is put side by side on a glass. If the sample shows weaker than the standard it is weighed out again with the same amount of zinc white and a little more color and the rubbing is repeated. This is done until the sample, with the color added, exactly matches the standard. Thus, if for .2 gram of the standard it takes .25 grams of the sample to make a match, the sample is called 5 per cent weak; if, on the other hand, the sample being tested shows stronger than the standard, zinc white is added to reduce the color to a match and the sample is said to be strong.

The authors have failed to get satisfactory results by using a greater amount of white as the difference in the strength is not so apparent under these circumstances as when a greater amount of color is used as recommended above. The use of chrome green or prussian blue with yellows has also been found not to give as satisfactory and uniform results as the use of white.

Where colors differ greatly in hue, judgment must be used in estimating their weakness or strength as it is impossible to get two pigments of the same color that differ in hue to look alike, and often this variation in hue is taken for weakness in color strength.

Fineness. — Fineness can be tested by the use of sieves of very fine mesh or bolting cloth, but as printing inks require colors of a great degree of fineness it is more satisfactory to judge the fineness by rubbing the pigment under the finger on a smooth piece of paper. After a little experience the fineness of a pigment can be easily judged by this method.

In some cases the microscope is of value in determining fineness.

Top Hue and Under Hue. — Weigh out one gram each of the sample and standard and add the same amount of oil to each, mix with a palette knife and rub each 50 times with a muller. Compare the top hues on glass. Then take some of the material and rub it out on a piece of paper side by side with the standard. Where the color is rubbed out thin the under hue will show up.

Bleeding in Oil. — Allow the paper with the rub-outs for under hue to hang for a day and if the oil that separates out is colored, the pigment is soluble in the oil and will bleed. This bleeding should be compared with the standard.

A pigment that bleeds in oil to any great extent should not be used in printing inks as it will strike through the paper and stain the other side, penetrate through a color laid over it or spread from the lines and give the work a dirty look.

Bleeding in Water. — Weigh one gram of the sample into an Erlenmeyer flask, shake up with 100 c.c. of distilled water, filter through a filter paper into a test tube and if the filtrate is colored the material bleeds in water.

The degree of this bleeding should not be very great as it shows either that the coloring matter in the lake is soluble or that it is imperfectly fixed on the base, and this condition is apt to cause trouble in classes of work where the paper is used wet or is liable to come in contact with water.

PART TWO

MANUFACTURE AND PROPERTIES OF INK-MAKING MATERIALS

SECTION ONE. DRY COLORS

1. REDS

VERMILLION

VERMILLION is the red sulphide of mercury HgS. There are two forms of HgS, the black sulphide, which is amorphous and the red sulphide which is crystalline. Vermillion, under the name of "Cinnabar" occurs in a great many places but natural vermillion, at present, is not found on the market as a pigment, as the artificial vermillion is its superior in color and brightness.

The manufactured vermillion on the market today is mostly made by the following process:

When metallic mercury is mixed with a concentrated solution of potassium pentasulphide at a moderate temperature a dark red powder is obtained, which is converted by a concentrated solution of caustic potash into bright red vermillion. As the color approaches bright redness great care must be taken as too high a temperature will produce a dull red of brown hue. The material is then washed with dilute caustic potash and afterwards with water, until the alkaline reaction disappears.

PROPERTIES OF VERMILLION

Top Hue	Under Hue	Fineness
Bright red.	Red of orange hue.	Not very fine. Lightens on grinding.
Flow	**Incompatibility**	**Abrasive Qualities**
Being heavy bodied it does not make an ink that flows well.	Cannot be used with pigments containing lead.	Not abrasive.
Oil Absorption	**Bleeding**	**Fastness to Light**
Requires only a small quantity of oil.	Does not bleed. When toned with dye toners will bleed in alcohol and other solvents.	Fast to light unless toned with dye toners.
Shortness	**Drying**	**Smoothness**
Somewhat short.	Poor drying pigment. Will rub off even after a long time.	Does not make a smooth ink particularly for typographic work.
Behavior towards Vehicles	**Atmospheric Influences**	
Works up well but separates on standing. It also has a tendency to stand up from the vehicle when printed, the latter going into the paper.	Will turn brown after exposure to air. The most widely held theory is that it gradually reverts to the amorphous condition, that is to the black sulphide, although this theory is by no means proven. This seems however to be substantiated by the fact that the red brown-hued pigment is produced when too much heat is used. The same hue is produced when an ink containing vermillion is subjected to prolonged grinding. The chemical composition of the two forms is the same.	

PROPERTIES OF VERMILLION — Continued

Value as an ink-making pigment	Vermillion is not a very good pigment to use in printing inks. It is high priced and does not work well as a typographic ink for the following reasons: it prints out unevenly on account of its shortness and being a poor drier it is liable to rub off and offset. Being a heavy pigment it separates out of the vehicle requiring remixing after standing a little while. For plate printing it works fairly well but is not of much value on account of its rubbing off, due to lack of drying, separation on standing and its tendency to darken. There are many lake colors that can be used in all cases requiring vermillion of lower cost, and that will give better working results for all classes of work.

2. BLUES

Ferrocyanide Blues

This type of blue comprises those pigments known as prussian blue, chinese blue and bronze blue. They are all of practically the same chemical composition, namely ferric-ferrocyanide [$Fe_4(FeCN_6)_3$]. This formula is however only theoretical as the commercial pigment varies according to the method of manufacture and it is extremely doubtful if it is ever of the above composition exactly.

Broadly the manufacture of these blues depends on the mixture of a ferrous salt, usually ferrous sulphate (copperas), with potassium ferrocyanide (yellow prussiate of potash) and the oxidation of the ferrous iron to ferric iron by a powerful oxidizing agent. These three varieties of ferrocyanide blues, while they have the same approximate chemical composition, differ from each other slightly in this and to somewhat a greater extent in physical characteristics. This variation is due to differences in the formula brought about by differences in the manufacture. These modifications will therefore be taken up in order.

A. Prussian Blue. — This is the darkest variety of ferrocyanide blue and has a decided red hue. For making a soft-drying prussian blue of good red hue the following process will be found satisfactory.

Solution No. 1	468 lbs.	Potassium Ferrocyanide
2	344 "	Ferrous Sulphate
		(dissolved in the same amount of water)
3	110 lbs.	66° Bé Sulphuric Acid
4	850 "	35° Bé Commercial Nitrate of Iron
5	44 "	Sulphate of Aluminum in water
6	112 "	Sodium Carbonate in water

Heat solutions No. 1 and No. 2 to boiling and run simultaneously into a tub of boiling water with constant stirring. When precipitation is complete add solutions No. 3 and No. 4, boil until oxidation is complete, wash free from iron and add No. 5 followed by No. 6; filter without washing and dry at 160° F. The filtrate contains about 25 pounds of blue as sodium ferrocyanide, which can be obtained by acidifying, precipitating with ferrous sulphate and oxidizing with commercial nitrate of iron. This gives a good pigment at moderate cost.

B. Chinese Blue. — This is a light variety of ferrocyanide blue with somewhat of a green hue. Its method of manufacture is similar to prussian blue as described before, only it is oxidized with hydrochloric acid and potassium chlorate without the final addition of alkali.

A bright chinese blue is produced as follows:

Solution No. 1	70	lbs.	Yellow Prussiate of Potash
	17	"	Hydrochloric Acid 20° Bé
2	45	"	Ferrous Chloride 35° Bé
3	$10\frac{1}{8}$	"	Hydrochloric Acid 20° Bé
	$10\frac{1}{3}$	"	Potassium Chlorate

Boil solution No. 1 till it is neutral and then run it and the 45 pounds of solution No. 2 heated to boiling into a tub of boiling water adding simultaneously solution No. 3, boil with steam until oxidation is complete, wash with water and filter.

C. **Bronze Blue.** — This is a variety of ferrocyanide blue that shows a metallic luster on drying. It can be made by the following formula:

Solution No. 1	400	lbs.	Yellow Prussiate of Potash in water
2	400	"	Ferrous Sulphate in water
3	136	"	36° Bé Nitric Acid
4	144	"	66° Bé Sulphuric Acid

Heat solutions No. 1 and No. 2 to boiling and run them simultaneously into a tub of boiling water; run in solutions No. 3 and No. 4, boil the entire solution till oxidation is complete, wash once and filter. This makes a brilliant blue of strong bronze hue.

Ultramarine Blue

Ultramarine blue is made from a mixture of aluminum silicate, sodium carbonate, sodium sulphate and charcoal. This is calcined in a crucible and ground. The proportions vary as regards the hue to be produced and it is very difficult to match a given hue. As the formula for a given hue depends on the composition of the raw materials no attempt to give formulas will be made.

PROPERTIES OF THE BLUES

Name	Top Hue	Under Hue	Fineness
Ultramarine blue.	Fairly dark blue.	Fairly light blue of red hue.	Very fine impalpable powder.
Prussian blue.	Dark blue.	Blue of red hue.	The ferrocyanide blues can be either soft or hard depending on the method of manufacture and the process of drying. The best of these blues are the ones that have been dried soft and feel velvety to the touch.
Chinese blue.	Dark blue.	Blue of green hue.	
Bronze blue.	Dark blue bronze tone in light.	Blue of green hue.	

Name	Drying	Shortness	Flow
Ultramarine blue.	Does not dry well. Has a tendency to rub off.	Is not short.	Flows well. Rather long in litho-varnish.
Prussian blue.	Ferrocyanide blues dry well. They exert a slight drying action themselves but not enough to make them harden in the fountain or the package.	Fairly short.	All the ferrocyanide blues flow well.
Chinese blue.		Not short.	
Bronze blue.		Not short.	

Name	Oil Absorption	Smoothness	Fastness to light
Ultramarine blue.	Rather low.	Makes a smooth ink.	Fast to light.
Prussian blue.	Good.	All ferrocyanide blues make smooth inks.	All the ferrocyanide blues are fast to light.
Bronze blue.	Good.		
Chinese blue.	Good.		

PROPERTIES OF THE BLUES — Continued

Name	Atmospheric Influences	Incompatibility	Value as an Ink-making Pigment
Ultramarine blue.	Darkens somewhat on exposure.	Cannot be mixed with lead colors.	Fair on account of its poor working qualities on the press and its lack of distribution. Makes a tacky ink with varnish and has a tendency to print out unevenly. Rubs off somewhat after drying.
Prussian blue. Chinese blue. Bronze blue.	All the ferrocyanide blues turn green on exposure to the atmosphere due to oxidation.	The ferrocyanide blues should not be mixed with paris white as they decompose it. For this reason they should be mixed on a straight barytes base.	With the exception of their change of hue on exposure these blues make first-class printing inks and are especially good for toning blacks. On account of price and their deep blue color there are no pigments that can really take their place.

3. YELLOWS

CHROME YELLOWS

These pigments, without doubt the most important and most used mineral pigments, are divided into three classes although they do vary from a light canary yellow through every hue of this color to a red of orange hue. It is customary, however, to class all chromate of lead yellows as either chrome yellows, chrome yellow of orange hue or chrome yellow of red hue and to consider the different modifications as hues of these three.

In this chapter the first-class or chrome yellows will be taken up, the other varieties being treated under the head of oranges. The chrome yellows are all essentially the normal chromate of lead of the formula $PbCrO_4$, the different hues being produced by different processes of manufacture, by which varying amounts of lead sulphate or carbonate are introduced and different conditions of physical structure are produced. Thus the more lead sulphate or carbonate present in the finished product the lighter the pigment produced will be. Different effects will also be produced depending on the size and character of the crystallization. The more crystalline the product is the darker it will be and the less will be its color strength.

Beginning with chrome yellow, which is the normal chromate of lead, $PbCrO_4$, we will therefore find a number of chrome yellows of slightly different hue, growing lighter as the amount of simultaneously precipitated lead sulphate or unused lead carbonate where white lead is used as a source of lead, increases.

In the manufacture of these chrome yellows care must be taken to precipitate the color in as nearly the amorphous condition as possible to insure the proper amount of color strength. This can be done by precipitating the pigment in a very dilute, cold, solution, with constant stirring. The washing should be done as rapidly as possible and the lead and chromate solutions should be run into the tub simultaneously. In order to exactly duplicate a hue, care should be taken that the same strength of solution, amount of water, stirring and temperature is used every time. If this is attended to the results of many different batches will duplicate each other very closely.

Chrome yellows of different hues are found that contain white lead instead of sulphate of lead or mixtures of white lead and sulphate of lead but the authors have found after a long series of tests that the chrome yellows which contain lead sulphate show the least darkening on exposure and have better working qualities.

A. **Chrome Yellow.** — This is the darkest form of the chrome yellows. It is the normal chromate of lead and has a slight orange hue.

For the production of a chrome yellow of a good deep yellow with a slight orange hue and of good color strength and of great softness the following formula will be found satisfactory:

Solution No. 1	100 lbs.	Lead Nitrate to 600 lbs. water @ 50° F.
2	35 "	Sodium Bichromate to 600 lbs. water @ 50° F.
3	2000 "	Water @ 50° F.

Run No. 1 and No. 2 simultaneously into No. 3 with constant stirring, let settle, wash and dry in vacuum or at about 160° F. As is seen from the above formula the lead should be somewhat in excess.

Between the chrome yellow mentioned above and chrome yellow of orange hue there are a variety of yellows of increasing red hue that are produced by the action of sodium or potassium bichromate and monochromate on basic lead acetate. The general way of producing this effect, which is the production of a certain small amount of basic lead chromate along with the normal lead chromate, is to neutralize the bichromate with an alkaline carbonate; thus converting part of the salt into the monochromate. These colors being mixtures of basic lead chromate and normal lead chromate should not rightfully be classed as chrome yellows, but they are not red enough in hue to be considered as oranges to which class by chemical composition they really belong.

B. Chrome Yellow Lemon. — This is the lightest of the chrome yellows and has a slight green hue. It is variously referred to as primrose yellow, lemon yellow and canary yellow, depending on the lightness of the hue. It is a mixture of normal lead chromate and sulphate or carbonate of lead. The chrome yellow lemon made from white lead is not as stable to light and has not the working qualities for printing inks that the straight sulphochromates have and it is preferable in the ink-making business to have a light yellow made with the sulphate alone.

A formula for making a lemon hued yellow of good color strength, brillancy and fastness to both light and atmosphere is as follows:

Solution No. 1 100 lbs. Lead Nitrate to 600 lbs. water at 50° F.
2 25 " Sodium Bichromate to 250 lbs. water @ 50° F.
3 35 " Sodium Sulphate to 250 lbs. water @ 50° F.
4 Mix solutions No. 2 and No. 3
5 2000 lbs. water at 50° F.

Run solutions No. 1 and No. 4 simultaneously, with constant stirring, into solution No. 5, wash well with water and dry at about 100° F.

As stated before a whole series of hues can be produced between chrome yellow and chrome yellow lemon hue by varying the amounts of bichromate and sodium sulphate or lead carbonate.

All lemon hued yellows have a tendency to darken in hue after being made. This difficulty can be overcome by adding a slight excess of sodium carbonate over the amount necessary to neutralize the acid set free in the above process and taking up any excess of soda by the addition of calcium chloride or by neutralizing the acid with freshly precipitated hydrate of alumina. For the above formula 7 pounds of calcium chloride 40° Bé and 12 pounds sodium carbonate added immediately after the precipitation of the chromate will be found satisfactory. This method of procedure explains why small amounts of calcium carbonate or alumina are found in a great many chrome yellow lemons.

The lightest chrome yellow lemons, the so-called sulphur or canary yellows, that have a very green under hue are produced by adding citric or tartaric acid to the bichromate solution along with the sodium sulphate. Some color makers hold the theory that the yellow pigments thus made will not darken on exposure to sunlight, but we have found that this is not the case, in fact pigments made with the citrate or tartrate of lead showed more darkening on exposure than those made with the straight sulphate.

The reason why a chrome yellow made in a very cold solution or with an excess of lead sulphate should stand the light better than any other form of the same chemical composition is not at the present time clear to us and we do not think it advisable to advance a theory in this regard so we will content ourselves with stating a fact that has come under our observation and leave the solution to another time.

PROPERTIES OF THE CHROME YELLOWS

Normal Lead Chromate and mixtures of Normal Lead Chromate with Lead Sulphate or White Lead and Basic Lead Chromate

Name	Top Hue	Under Hue	Fineness	Incompatibility
Chrome yellow lemon hue.	Lemon yellow.	Slightly green.		
(Normal lead chromate with lead sulphate or white lead.)	There are a number of different top hues under this class from a very light sulphur color to a moderately dark lemon.	All of this class of yellows are distinguished by a more or less green under hue.		
Chrome yellow.				
(Normal chromate of lead.)	Deep yellow, slightly orange.	Slightly red.		
Other modifications between the normal lead sulphate and the lead sulpho-chromates or mixtures of normal lead chromate and white lead.	There are a number of hues between chrome yellow of lemon hue and chrome yellow getting deeper and more orange as they near the normal lead chromate. The middle hue of these is called chrome yellow medium.	As these colors near chrome yellow they pass from green to yellow and then to red in under hue.	Good soft colors.	Cannot be mixed with sulphur colors or used in the presence of alkalies.
Mixtures of small amounts of basic lead chromates with normal lead chromate.	There are a number of hues between chrome yellow and chrome yellow of orange hue that are not quite oranges but which are more red than chrome yellow.	There are a number of hues between chrome yellow and chrome yellow orange hue that are progressively more red as the amount of basic lead chromate increases.		

PROPERTIES OF THE CHROME YELLOWS — Continued

Name	Abrasive Qualities	Drying	Shortness	Flow	Oil Absorption
Chrome yellow lemon hue. (Normal lead chromate with lead sulphate or white lead.)					
Chrome yellow (Normal chromate of lead.)					
Other modifications between the normal lead sulphate and the lead sulphochromates or mixtures of normal lead chromate and white lead.	Do not exert any abrasive qualities.	Dry well. Not too fast. Drying increases as the composition of the yellow nears the normal lead chromate.	Have good length in oil or varnish.	Have good flow.	Low oil absorption. Work up well in vehicles.
Mixtures of small amounts of basic lead chromates with normal lead chromate.					

PROPERTIES OF THE CHROME YELLOWS — Continued

Name	Smoothness	Fastness to Light	Atmospheric Influences	Value as an Ink-making Pigment
Chrome yellow. Lemon hue. (Normal lead chromate with lead sulphate or white lead.) Chrome yellow. (Normal chromate of lead.) Other modifications between the normal lead sulphate and the lead sulpho-chromates or mixtures of normal lead chromate and white lead. Mixtures of small amounts of basic lead chromates with normal lead chromate.	Make smooth inks.	Not very fast to light. Darken somewhat on exposure. Those colors made with lead sulphate darken less than those without sulphate. Normal lead chromates do not darken much if they are pure products.	Sulphur gases in the air affect all chrome yellows.	Very good on account of the variety of hues and their working qualities. They are an important class of pigments for all kinds of printing inks.

4. GREENS

Chrome Green

While there are a number of different mineral greens on the market the only green of any importance to the ink maker is the so-called chrome green, which, in reality is a mechanical mixture of chrome yellow and prussian blue. This green can be prepared in two ways, wet and dry, but only that produced by the wet process is suitable for printing inks. There are a great many hues of this pigment that can be attained by variations in the process of manufacture and the materials used, so many in fact, that nothing but a general description of the method of making this pigment can be attempted here.

Chrome green for use in printing inks should be made only from pure chrome yellow and prussian blue without the addition of barytes or any other base except lead sulphate. The chrome yellow should be precipitated first and washed several times and freshly made prussian blue run into it in the form of a thin paste, with constant stirring. The prussian blue best suited for making bright greens is a blue of slight reddish hue about half between a blue of green hue and a blue of decided red hue.

The different hues of chrome green are produced by varying hues of chrome yellow and different amounts of prussian blue. Thus, when mixed with the proper amount of blue, a chrome yellow of lemon hue will give a bright, light chrome green. While a chrome yellow (normal lead chromate) will give a very dark green. Between these two almost any hue can be produced by varying the amount of blue and using a lighter or darker yellow. Care should be taken not to employ a yellow of decided orange hue or a yellow with a red under hue as the resultant green will be an olive or a dirty green. The yellows for making chrome greens should always be the lightest and brightest that can be made.

Some firms have been very successful in making a bright, resistant green by the use of zinc yellow, that is, the chromate of zinc. It has been our experience, however, that this color does not give the working qualities that the lead chromate greens give.

Chromium Oxide

The oxide of chromium is really the best and fastest green that can be secured but it can only be obtained at its best in one hue. It is also a very expensive color, too expensive in fact for general use except as a tinting color or for the highest grades of work.

Properties of Chrome Greens

Name	Top Hue	Under Hue	Fineness	Incompatibility
Chrome green.	Varies in top hue from a light green to almost black.	Varies from a light green to a very dark blue.	Always fine.	Incompatible with sulphur colors and alkalies.
Chromium oxide.	Medium green.	Medium green	Fine.	Unaffected.

Name	Drying	Shortness	Flow	Oil Absorption
Chrome green.	Dries well.	Makes a long ink.	Flows well.	Low, mixes well with vehicles.
Chromium oxide.	Dries well.	Long in oils.	Flows well.	(Same as above.)

PROPERTIES OF CHROME GREENS — Continued

Name	Fastness to Light	Atmospheric Influence	Value as an Ink-making Pigment
Chrome green.	Fast to light.	Blues somewhat on exposure.	A very valuable pigment on account of its price and stability under all conditions except those of an alkaline nature.
Chromium oxide.	Absolutely fast to light.	Fast to all atmospheric influences.	An excellent pigment under all conditions but too expensive except for the highest kind of work.

5. ORANGES

CHROME YELLOW ORANGE

The chrome yellow oranges, which include all the hues of orange between chrome yellow orange hue and chrome yellow red hue are very important pigments. They are the best mineral oranges for printing-ink work. Chrome yellow orange hue is a mixture of normal lead chromate and basic lead chromate, while chrome yellow red hue is the basic chromate of lead. These hues, which have a very wide range, are produced by increasing the amount of basic lead chromate in the pigment until it finally consists entirely of the basic chromate, which is chrome yellow, red hue.

A. Chrome Yellow, Orange Hue.—This is a bright pigment of true orange hue. It is generally made by converting part of the normal lead chromate of the chrome yellows into basic lead chromate, by treatment with caustic soda. If normal lead chromate is treated directly with caustic soda there will be a loss of chromium, which goes off in the filtrate as sodium chromate. In the following method, which gives a bright, soft, pigment of decided orange hue and good color strength, the loss of chromate is avoided, as the sodium chromate formed in the decomposition of the normal lead chromate is used up by the lead sulphate. By varying the amount of sulphate and caustic soda any number of different degrees of orange hue can be obtained.

Solution No.	Amount		
1	100	lbs.	Lead Nitrate in 600 lbs. water @ 50° F.
2	18¾	"	Sodium Bichromate in 600 lbs. water @ 50° F.
3	12¼	"	Sulphuric Acid 66° Bé
4			Solutions No. 2 and No. 3 mixed
5	2000	"	Water at 50° F.
6			Caustic Soda 40° Bé

Run solutions No. 1 and No. 4 simultaneously into No. 5, decant off the top liquor, heat No. 5 to boiling and add No. 6 until the desired color is reached.

B. Chrome Yellow, Red Hue. — This pigment is basic chromate of lead. It has a dull orange red top hue and an under hue of orange with a slight red hue. It could be classed as a red but on account of its chemical composition and the decided orange of its under hue it is more properly classed as an orange.

The following formula will give a good type of this pigment:

Solution No.	Amount	
1	100 lbs.	White Lead, well ground in 30 lbs. of water.
2	31 "	Potassium Bichromate in 110 lbs. of water, this solution being neutralized with 281 lbs. of crystallized Soda.
3		No. 2 is heated to boiling until the Carbonic Acid is expelled and No. 1 is added as slowly as possible. The resultant precipitate is then drawn off into another tub and washed twice. Enough water is then added to make it up to the original volume of the mixed solutions.
4		Sulphuric Acid 66° Bé., about 4 lbs. to every calculated 100 lbs. of Basic Lead Chromate in No. 3.

Run No. 4 into No. 3 in the cold with constant stirring. The material is then washed once with hot water and dried.

Orange Mineral

Orange mineral is another pigment of orange hue; but its value as an ink-making pigment is not nearly so great as that of chrome orange, since it has several very serious physical defects. Orange mineral consists of PbO and PbO_2 chemically combined to make Pb_3O_4, in the following percentages, 35 per cent PbO_2 and 65 per cent PbO. This percentage varies a little in different samples. Three varieties of orange mineral are recognized in the color trade namely, German, French and American. They are all of approximately the same chemical combination, but differ greatly from each other in physical characteristics. This is due entirely to differences in the raw materials used to produce the pigment.

The most serious drawback to orange mineral is its tendency to form a soap with linseed oil and thus to produce a livery ink, and the fact that an ink made from it will harden so that it is unfit for use after standing for a short while. French orange mineral shows this defect least of the three varieties, while American orange mineral is on this account practically worthless for printing-ink work. All of the orange minerals settle out badly and show a great many defects in working qualities. The different varieties also have slightly different hues.

Methods of Manufacturing Orange Minerals. — Both French and German orange minerals are made by heating white lead in a muffle furnace until the carbonic acid and moisture are driven off and the resulting lead oxide converted into the proper proportions of dioxide and monoxide. French orange mineral is made by using white lead made according to the French method while German orange mineral is made from white lead made according to the German method. The American orange mineral, which is nothing but what is ordinarily called red lead of a slight orange hue, is made by heating litharge in a muffle furnace.

PROPERTIES OF THE ORANGES

Name.	Top Hue	Under Hue	Fineness	Fastness to Light	Incompatibility
Chrome yellow, orange hue	Bright orange	Yellow orange		Very Fast.	
Chrome yellow, red hue.	Dull orange red	Orange of decided yellow hue.		Very Fast.	
French orange mineral.	Bright orange of slight red hue.	Bright orange of slight red hue.	Impalpable powder.	Darkens slightly under light.	Affected by sulphur colors
German orange mineral.	Bright orange, slightly lighter than the French.	Bright orange with a slight red hue. More yellow than the French.		Darkens slightly under light.	
American orange mineral.	Orange of decided red hue.	Orange of decided red hue.		Darkens slightly under light.	

PROPERTIES OF THE ORANGES — Continued

Name	Atmospheric Influences	Abrasive Qualities	Oil Absorption	Drying	Shortness
Chrome yellow, orange hue.	Affected by sulphur gases.	Not abrasive.	Very low.	Exert considerable drying action. Care should be taken in using driers.	Is of good length.
Chrome yellow, red hue.				Exert considerable drying action. Care should be taken in using driers.	Is of good length.
French orange mineral.				Dries quickly.	Fairly long.
German orange mineral.				Dries very quickly.	Fairly short.
American orange mineral.				Dries very quickly.	Very short.

PROPERTIES OF THE ORANGES — Continued

Name	Flow	Smoothness	Value as an Ink-making Pigment
Chrome yellow, orange hue	Good.	Very smooth.	Very important. A fairly cheap pigment that works well in all kinds of ink.
Chrome yellow, red hue.	Good.		Used in making mixed or degraded colors where a red top hue and yellow orange under hue is desired. The only pigment of this character there is.
French orange mineral.	Flows fairly well.		Is not of very much value. But it is the best of the three kinds of orange mineral as it does not thicken as readily as the other two.
German orange mineral.	Flows fairly well.		Thickens up and livers on standing. Not of much value.
American orange mineral.	Does not flow very well.		Of no value as an ink-making pigment.

6. RUSSETS

All the inorganic pigments that can be classed as russets owe their hue primarily to iron oxide; the different names met with are either given to them as trade names or to differentiate between various hues. As a rule these iron oxide pigments, whether natural earth pigments or artifically made by precipitation of iron or calcining by-products, have practically no value as typographical ink pigments and very little value as plate ink pigments on account of their hardness, which causes abrasion of the plates in plate printing. Another drawback is the difficulty of getting them in a fine state of division, as any slight coarseness makes them fill up the forms in typographic work.

All of these colors, on account of their low price, are used on cheap poster work but as the same or even better effects can now be produced for high grade work by colors with superior working qualities their work is limited to the above field and to work where an alkaline condition is to be met with, such as wrappers for soap and labels for lye cans.

Indian Red

Indian red is a clay containing a high percentage of iron oxide. The general way of preparing it is to levigate the crude color to remove the sand and other impurities, calcine it in a furnace and regrind the calcined product. As can easily be imagined the resulting pigment lacks fineness of grain and is generally contaminated with fine sand.

Venetian Red

Venetian red is an iron oxide pigment made in two ways, either by precipitating a solution of ferrous sulphate with soda and calcining the precipitate or by calcining ferrous sulphate in a muffle and then recalcining the ferric oxide with salt. The hue varies with the amount of salt used. This pigment is a hard grained material seldom of any degree of fineness due to the difficulty of grinding it.

Burnt Umber

Burnt umber is the calcined form of a clay containing iron oxide and manganese, called raw umber. It is somewhat similar in physical characteristics to indian red and is treated in the same way. It is rarely very fine, frequently contains grit and sand, and is not of much value except for mixing a small quantity in an ink to produce a certain shade.

Burnt Sienna

Burnt sienna is the calcined product of raw sienna, an earth containing a large percentage of ferric hydroxide. It is similar in physical characteristics to the other iron pigments. Its value is about the same as that of umber.

7. CITRINES

The pigments that can be classed as citrines are the raw forms of sienna, umber and ochre. They also, as do the russets, owe their hues to varying amounts of iron and manganese oxides. They have little or no value as ink-making pigments. The method of preparing these pigments, which is practically the same for all three, consists in levigating, drying and grinding. Care should be taken in drying as these pigments change color at high temperatures.

Properties of Russets and Citrines

Name	Top Hue	Under Hue	Fineness	Incompatibility	Abrasive Qualities
Indian red.	Dull purple red.	Dull brown red.	Coarse.	Can be used with any color or base	Very abrasive.
Venetian red.	Light brown red.	Dull brown red.			
Burnt umber.	Dark brown.	Dull yellow brown.			
Burnt sienna.	Dark brown red.	Light red brown.			

PROPERTIES OF RUSSETS AND CITRINES — Continued

Name	Drying	Shortness	Oil Absorption	Flow	Fastness to Light
Indian red.	Dries poorly	Short.	Quite high.	Poor.	Fast.
Venetian red.			Quite high.		
Burnt umber.			Very high.		
Burnt sienna.			Very high.		

Name	Atmospheric Influences	Smoothness	Value as Ink-making Pigments
Indian red.	Not affected.	Very grainy.	Poor.
Venetian red.			
Burnt umber.			
Burnt sienna.			

8. BLACK PIGMENTS

From the point of view of wideness of use and quantity used the black pigments are without doubt the most important ink pigments. With but one exception they are all manufactured products and all but two of them owe their blackness to amorphous carbon. The black pigments are practically all good ink-making pigments, different ones being adapted to different classes and processes of printing; while one or two are valuable as adjuncts, that is as materials to add to certain inks to improve their working qualities.

The black pigments may be primarily divided into five classes as follows:

Bone Black.

Vine Black or Vegetable Black.

Carbon or Gas Black.

Lamp or Oil Black.

Miscellaneous Blacks (including natural blacks and certain blacks made by patented processes, from by-products or from mixtures of certain others, that do not fall readily into the above classifications).

A. Bone Black. — Bone blacks are, as the name implies, made by the calcination of bones in air-tight retorts. They consist essentially of carbon and an ash, consisting of calcium phosophate and calcium carbonate with a small amount of calcium sulphate and sulphide, these salts being the mineral constituents of the bones. This ash amounts to about eighty-five per cent of the black.

The bone blacks found on the market are of three different varieties: sugar-house bone black, that is bone black that has already been used to refine sugar; bone black primarily made for use as a pigment and acid washed bone black. The method of manufacturing the bone black is the same in all cases but on account of the after treatment, the different varieties have different physical properties.

Method of Making Bone Black. — The general process of making bone black is to select the densest bones, boil them to remove the fat, grind to a coarse powder and burn them in retorts from which the air is excluded by luting. In most bone-black factories the by-products of the dry distillation, ammonium carbonate, bone oil, etc., are recovered in a suitable manner. When the bones are thoroughly calcined the retorts are allowed to cool and the product withdrawn.

For use in refining sugar the material is left in its granular form. After being used for filtering sugar until its decolorizing power is entirely used up the black is washed, ground wet, dried, bolted and finds its way to the market as a pigment.

In preparing bone black for a pigment primarily, the dense hard bones are ground very fine, sifted and then burnt, the products of combustion in most cases being burnt instead of recovered. This, it is claimed, makes a better product than when the by-products are recovered, although it is impossible to see any reason why this should be so. As in many similar cases this appears to be a tradition with no practical foundation. The bone black prepared in this way has greater strength of color and better working qualities than sugar-house bone black.

When bone black is treated with acid the calcium salts are dissolved out and the result is a very finely grained carbon containing little or no ash, depending on the amount of acid treatment. Acid treated bone black has a very deep black color and in consequence of its fine state of division a great deal of color strength. It is used as a toner for slightly off-colored or weak bone blacks and as a forcing black to bring up the color on certain classes of plate work. As a toner for bone blacks it is far superior to lamp or carbon black on account of its lower oil absorption and its better working qualities.

The use of bone black is entirely restricted to plate printing inks. Bone black is also designated by the plate printing trade as hard black. It is sometimes put on the market in lumps or in the form of drops, the idea being that blacks in this form are purer and better than the powdered blacks; this is of course, a great fallacy as the pigments in both cases are identical.

B. Vegetable or Vine Black. — Under the head of vegetable or vine blacks are classed the black pigments produced by the dry distillation and carbonization of willow wood, wine yeast, grape husks, grape-vine twigs, spent tan bark, shells, fruit pits, sawdust or in fact any material of vegetable origin.

The blacks made from grape-vine twigs, grape husks and wine yeast are the best of this class and the name vine black is derived from the fact that they were formerly used exclusively to produce it. Of late years however this name has been extended to take in every variety of black of vegetable origin. These vegetable blacks, when properly made, are of a deep black color and show great strength. They consist of about 75 per cent of carbon and 25 per cent of ash. The ash contains silica, calcium phosphate, carbonate and sulphate and a small amount of potassium carbonate.

The vegetable blacks can be graded in the following manner:

True vine black, that is black from wine yeast or grape-vine twigs.

Blacks from grape husks.

Blacks from willow charcoal.

Blacks from spent tan bark.

Blacks from fruit pits.

The differences in quality are due entirely to the different physical properties exhibited by the products from different sources: such as differences in density, grain and state of division and to the process employed in carbonizing them and treatment after carbonization.

Method of Making Vine Black. — In the manufacture of vine black practically the same process is used as in the manufacture of bone blacks, that is the raw materials are carbonized in the absence of air. This is done in a suitable retort, a number of different forms of apparatus being used, with or without the recovery of by-products.

After the material is burnt, it is well washed with water to remove the soluble alkaline salts and ground. The very finest grades of vine black are also acid washed, which produces about the same effect as it does in the case of bone blacks. In the plate printing trade in which trade they are only used, they are called soft blacks. Their principal use is to give a certain grain to bone blacks and to give increased color.

C. **Carbon or Gas Black.** — Carbon or gas blacks are almost pure carbon containing practically no ash. They consist of about 92 to 95 per cent of carbon, the remainder being moisture and unburnt hydrocarbons from the raw material. The amounts of moisture and unburnt hydrocarbons are of course variable with the process of manufacture and the care taken in collecting the finished product.

The carbon blacks are very fine pigments of low specific gravity and are consequently very bulky. They are a very deep black and make excellent typographical inks of fine working qualities. They cannot, however, be used in plate inks on account of their high oil absorption and lack of body which makes them hard to wipe or polish.

There is a wide range in the behavior of different carbon blacks when made into inks, a wider range than there is any apparent cause for. In working qualities, the different brands and grades of carbon black on the market exhibit this to a great extent. They vary from making short, tacky inks to making inks of great natural length and decided flow while some carbon blacks which we have had experience with, were absolutely worthless for making any kind of ink, even news ink. A great difference will also be found in the behavior of different carbon blacks towards driers. The authors have done a great deal of work along these lines but have been unable to find any reason as to why the working qualities and behavior of blacks of the same general origin and composition should vary in their physical properties to such an extent. By the proper combination of varnishes and grades of black almost any desired effect can be produced in typographical printing ink by the use of this pigment.

Manufacture of Carbon Black. — There are numerous different methods of making gas black which are all similar in the essentials. That of burning natural or some form of producer gas with a small supply of air and collecting the soot in a suitable way is the most common. The methods of burning the gas and of collecting the resultant pigment vary greatly, a number of patents having been granted both in this country and abroad on special apparatus for burning and collecting carbon black.

The apparatus most commonly in use in this country at the present day consists of a burner, generally ordinary pipe, but in some cases a regular burner of the Argand type, and a water-cooled surface for collecting the soot. This water-cooled surface is rotated above the burner and is provided with a means for removing the soot as it accumulates. The carbon black made in this way is a soft, bulky, deep black product and contains only a small amount of unburnt hydrocarbons.

D. **Lampblack.** — Lampblacks like carbon blacks are almost pure carbon and are similar in appearance to carbon blacks with the exception that, as a rule, they are not in such a fine state of division and contain more products of distillation. A very fine calcined lampblack, that is one which has been reheated after making, shows about the same properties as carbon black. As is the case in carbon blacks, there are a great many grades of lampblack on the market, depending on the raw materials used, the methods of manufacture and on the state of division of the product. Each of these grades shows certain different physical properties.

A number of raw materials have been and still are used for making lampblack, among which, may be noted, naphthalene, anthracene, oil residues, pitch, resin and gas tar oils, the latter being the most widely used at present. Nearly all of the lampblack now on the market is made from the "dead oil" of the gas house.

There are a number of patents relating to apparatus for burning and collecting lampblack, the process being fundamentally the same as that used in making carbon black. There is, however, a difference in manipulation and in the details of the processes.

The most used method of producing lampblack is as follows: the material to be burnt is introduced into a cast-iron furnace, fitted with a door for the purpose and containing a pan or receptacle which has been heated to redness. This ignites the material to be burnt and the products of combustion are led into collecting chambers. The furnace is recharged from time to time by ladles, care being taken not to allow too much air to enter the furnace during the charging. The heating pans must be changed and cleaned every few days.

The collecting chambers consist of flues with staggered walls dividing them into compartments. These walls act in a way as baffle plates to stop the passage of the soot particles. The heavier particles will be deposited in the first compartment, while the finer ones are carried through and are deposited at the very end of the collecting chambers. The lampblack deposited in the first compartment is large grained and contains quite a large percentage of unburnt hydrocarbons while that collected in the last compartments is in a very fine state of division and is almost pure carbon. In some cases the lampblack collected in these chambers is recalcined in steel drums to rid it of unburnt hydrocarbons and the resultant product is of almost as fine a grade as carbon black. Of late years carbon black has almost entirely taken the place of lampblack in printing ink.

E. Miscellaneous Blacks

A. Mineral Black. — This pigment is a clay shale containing about thirty per cent of carbon. When properly prepared, that is washed and ground, it forms a fine, soft powder. It is sometimes used in making mixed blacks for plate printing inks.

B. Manganese Black. — The precipitated dioxide of manganese makes a brownish black that is of value when mixed with bone and vine blacks in the preparation of plate printing ink. The admixture of a small amount of this material makes a plate printing ink made from bone and vine black and prussian blue or from a mixed black and prussian blue work better, and reduces the amount of gathering.

C. **Magnetic Pigment.** — Under the name of magnetic pigment a very finely divided black oxide of iron, produced by a patented process, is put on the market. It is useful in black plate inks to give greater density, smoothness, and better working qualities. This material also has a tendency to prevent gathering.

D. **Special Blacks.** — Blacks for various special purposes, particularly for adding to other materials, are made from coke, lignite and certain by-products. The ash and carbon contents vary of course with the raw material.

E. **Mixed Blacks.** These are a class of blacks made up from bone blacks of different kinds, vine blacks and combinations of the different miscellaneous blacks mentioned above with the view of overcoming the defects of straight mixtures of bone and vine blacks. They contain between 30 and 50 per cent of ash, not over 5 per cent of which should be insoluble in hydrochloric acid, the remainder being carbon. These mixed blacks should never contain carbon black or lampblack as these materials are too hard to wipe and have too high an oil absorption for use in plate inks for which mixed blacks are only used.

PROPERTIES OF BLACKS

Name	Top Hue	Under Hue	Oil Absorption	Fineness
Bone black.	Greenish black.	Brownish black.	Fairly low.	Should be fairly fine, that is, they should still have some grain.
Vine black.	Greenish black darker than bone black.	Brownish black.	Fairly low.	
Carbon black.	Deep black.	Brownish black.	High.	
Lampblack.	Deep black.	Brownish black.	High.	
Mineral black.	Brownish black.	Decided brown.	Fairly low.	Should be impalpable powders.
Magnetic pigment.	Brownish black.	Decided brown.	Low.	
Manganese black.	Brownish black.	Decided brown.	Low.	

Name	Flow	Shortness	Fastness to Light	Atmospheric Influences
Bone black.	Flows fairly well.	Fairly short.		
Vine black.	Flows fairly well.	Fairly long.		
Carbon black.	Poor.	Short.		
Lampblack.	Poor.	Short.		
Mineral black.	Good.	Fairly long.	No effect.	No effect.
Magnetic pigment.	Good.	Long.		
Manganese black.	Good.	Long.		

PROPERTIES OF BLACKS — Continued

Name	Drying	Smoothness	Abrasive Qualities	Incompatibility
Bone black.	Exert no drying action.	Does not make a smooth ink.	Quite abrasive.	Mixes with everything.
Vine black.		Does not make a smooth ink.	Quite abrasive.	
Carbon black.		Works up very smooth.	Not abrasive.	
Lampblack.		Works up smooth.	Not abrasive.	
Mineral black.		Does not make a smooth ink.	Quite abrasive.	
Magnetic pigment.		Works up very smooth.	Not abrasive.	
Manganese black.		Works up very smooth.	Not abrasive.	

Name	Value as an Ink-making Pigment
Bone black.	Of great value as a plate printing ink material although it must be mixed with vine black for color and to give the proper working qualities.
Vine black.	Of great value as a toner to mix with bone black to give color and working qualities to black plate printing inks.
Carbon black.	The most important typographical black, in fact it is the base of all black typographical inks at the present day.
Lampblack.	Not much used at present as its place has been taken by the cheaper but similar carbon black.
Mineral black.	Used principally to mix with other blacks.
Magnetic pigment.	Used principally to mix with other blacks.
Manganese black.	Used principally to mix with other blacks.

The mixed blacks, being mixtures of various amounts of the above pigments, will have properties that vary according to properties and amounts of the different ingredients used in their manufacture.

9. DILUTENTS

Under the head of dilutents are classed those pigments that can be used to produce tints of the other pigments. They are of necessity white pigments of great covering power and strength. The principle dilutents used in printing ink manufacture are, zinc white, white lead and lithopone. Of these dilutents the most satisfactory, working qualities and brightness of the tint considered, is zinc white; lithopone comes next, but white lead is a very unsatisfactory tinting pigment. These white pigments are also used to make printing inks for special purposes, particularly for use on tin where a white background is necessary for color printing.

Zinc White

Zinc white is zinc oxide of the formula ZnO. It is a very soft pigment of a good body and covering power. The process of manufacture consists in roasting either metallic zinc or zinc ore till fumes are given off, burning these fumes in an atmosphere of super-heated air and carbon monoxide which accelerates the oxidation of the zinc and condensing the resulting oxide in suitable chambers. As is the case in the preparation of lampblack, the zinc oxide deposited first is coarse and heavy and frequently contains particles of metallic zinc, while that deposited farthest away from the retort or rotary furnace is the lightest and finest grade. The different raw materials and the care with which the process is carried out produce the different grades of zinc oxide.

Zinc oxide can be mixed with any other pigment and is not influenced either by light or atmospheric conditions.

Lithopone

Lithopone is a mixture of artificial barium sulphate and zinc sulphide. It is a soft white pigment of good covering power, in fact, its covering power is very little inferior to white lead and it can be used with excellent results to form tints, although not quite as valuable in this respect as zinc white, which has superior working qualities.

Method of Making Lithopone. — The following is the general method of manufacturing lithopone. Natural barytes is thoroughly mixed with finely ground lignite in the proportion of about 100 pounds barytes to 38 pounds lignite. This mixing is best done by grinding the two materials together in the wet way, as the yield of barium sulphide depends to a great extent on the intimacy of this mixture. The mixture is then dried and put into fireclay retorts with air-tight covers, the retorts themselves being provided with suitable outlets for the products of combustion. These retorts, which are generally arranged in batteries, are heated from the outside. The reduction of barium sulphate takes about four hours and the material in the retorts should be turned after about half this time. The turning should be done rapidly to prevent reoxidation of the barium sulphide which takes place very readily at high temperatures. When the reduction is complete the barium sulphide is dumped into air-tight iron boxes to cool. Care should be taken to have as pure barytes as possible, as impurities will have an injurious effect on the resulting lithopone.

The barium sulphide is then dissolved in water and a solution of zinc sulphate free from iron is made. These two solutions, the strength of each being known, are run into each other in approximately molecular proportions. There should, however, be an excess of zinc sulphate. The precipitate is washed and dried, calcined in retorts until cherry red and quenched suddenly in cold water, after which it is ground in wet mills for a long time, pressed, dried and reground dry.

Lithopone varies in the amount of zinc sulphide, the larger the amount the better will be the product. Brands of lithopone on the market show a variation in the amount of zinc sulphide from 15 to 33 per cent.

White Lead

White lead is a mixture of basic lead carbonate and lead hydrate with the formula $2\ PbCO_3PbH_2O_2$. As white lead is of little value as an ink-making pigment and the general methods of its production are well known, a brief explanation of the differences between the French and German methods of preparation is all that will be given here. The German process depends on the corrosion of metallic lead by acetic acid, the action being accelerated by heat, and the conversion of the basic lead acetate into lead carbonate by the introduction of carbonic acid gas into the acetic acid vapors. The French process consists in treating a solution of basic lead acetate with carbon dioxide.

French and German white leads show different characteristics; that made by the German process, showing more covering power but being inferior in whiteness and brillancy to the French. German white lead also has less oil absorption. These differences are partly due to physical structure and partially to varying proportions of lead hydroxide and carbonate.

PROPERTIES OF DILUTENTS

Name	Fineness	Covering Power	Strength	Incompatibility
White lead.		Very great.	Great.	Cannot be used with sulphur colors.
Lithopone.	Impalpable powder.	Fair.	Fair.	Cannot be used with lead colors.
Zinc white.		Good.	Good.	Can be used with any color.

Name	Fastness to Light	Atmospheric Influences	Oil Absorption	Flow
White lead.	Yellows on exposure to light.	Blackens in air due to sulphur gases.	Low.	Good.
Lithopone.	Darkens on exposure to light.	Affected somewhat.	Fairly low.	Good.
Zinc white.	Does not change.	Not affected.	Low.	Fairly good.

PROPERTIES OF DILUTENTS — Continued

Name	Shortness	Smoothness	Abrasive Qualities	Value as a Dilutent
White lead	Fairly long.	Makes a smooth ink.		Not of much value. Chalks through when used as a dilutent.
Litho-pone.	Fairly long.	Makes a smooth ink.	Not abrasive.	Can be used in both plate and typographic ink. Mixtures of aluminum hydrate and zinc white give better results in typographic work.
Zinc white	Quite long.	Very smooth.		Of exceptional value for making tints and on account of its working qualities in both plate and typographic inks. The best of the white pigments.

10. BASES

In the manufacture of printing inks there is a class of pigments that are of no value as pigments when used by themselves, but form a very important group when used with other pigments. These pigments are what are known as bases. That is, they are used to give body to and to carry other pigments. The bases have very little or no coloring power of their own but impart to the pigments they are used with, working qualities and body. They are in a few instances also used to produce tints.

The bases are used for a number of different purposes and to produce a number of different results as is shown by the following examples: for instance, in cases where a pigment shows a high oil absorption the mixture of a base of low oil absorption will often overcome this defect. A certain amount of base added to a pigment which in itself is quick drying will sometimes make an ink of moderate drying qualities. In plate printing work the defect of striking through, gathering, softness and stickiness can often be overcome by the use of a proper base. The bases are also extensively used as carriers for the organic dyes in the manufacture of lakes.

The principal bases are barytes (natural barium sulphate) blanc fixe (artifical barium sulphate), paris white (calcium carbonate), aluminum hydrate, magnesium carbonate and calcium sulphate. Of these aluminum hydrate, blanc fixe and magnesium carbonate are the only ones which are of value in typographical inks.

Barytes

The natural sulphate of barium is found more or less purely in almost all parts of the world. It is mined, separated into classes according to its whiteness and ground. Barytes requires a great deal of grinding to reduce it to the proper fineness and the best varieties are water floated after grinding. In a great many instances a small amount of ultramarine blue is added to the product to heighten its whiteness. This is frequently done when the material is contaminated with iron, the blue destroying the red or yellow hue imparted by the iron.

Barytes is a heavy white mineral of a specific gravity between 3.9 and 4.5. It is absolutely indifferent to all other chemicals, even acids, and can only be decomposed by fusion with alkalies. It has very little covering power and can, therefore, be added to other pigments without affecting their strength, and, having a low oil absorption, it is of great value when mixed with pigments which absorb a great deal of oil or varnish. On account of its density and grain, barytes, when mixed with pigments for plate inks, overcomes a great many of the defects in working qualities inherent in many mineral pigments; particularly is this true of the chrome colors which are of little value as plate inks by themselves.

For fine work the barytes should always be added in the manufacture of the ink itself and not as an adulterant in the dry color. Water floated barytes is superior for plate inks to blanc fixe because of its grain which makes the ink wipe clean and polish well. Very finely ground barytes is also of great value as a base for organic lakes. The character of the dye and the method of precipitation exert a great influence on the value of barytes as a lake carrier.

Paris White

Paris white is practically pure calcium carbonate. It is prepared by levigating chalk with water to remove grit and other physical impurities. Chalk, being an amorphous form of calcium carbonate, makes a far better base than ground limestone.

Paris white is a soft pigment with a specific gravity of about 2.5. It has very little covering power and dries quite slowly. It is generally mixed with barytes in plate inks to improve the polishing and is used to some extent in making organic lakes.

Aluminum Hydrate

Aluminum hydrate is a very important base both for printing inks and organic lakes. When added to plate inks it makes them work better, particularly in the polishing. In typographic inks it increases the length and flow, correcting any tendency to shortness and unequal distribution. It is also used with good effect in inks that dry too much themselves, acting in these cases like a reducer. Inks that do not grind up smooth are also helped by the addition of aluminum hydrate.

Aluminum hydrate has the formula $Al_2(OH)_6$, but most of the commercial aluminum hydrate consists of a mixture of basic aluminum sulphate and aluminum hydrate. This condition arises from the fact that basic aluminum sulphate is not very readily converted by sodium carbonate into aluminum hydrate in cold solutions and it is necessary to precipitate the hydrate in cold solutions in order to get a soft and powdery product on drying. Aluminum hydrate precipitated by alkaline carbonates in hot solutions dries hard and horny. Therefore, when precipitation takes place in the cold a part of the basic aluminum sulphate formed in the process is left unconverted.

The method of manufacturing aluminum sulphate is as follows:

The solution of aluminum sulphate is treated with an excess of sodium carbonate with rapid stirring until the carbonic acid ceases to be evolved. It is then washed and dried. This process as stated above gives a superior product when dried than when the solution is boiled.

Blanc Fixe

Blanc fixe, which is the artificial sulphate of barium, is prepared from barytes by calcining finely ground barytes with powdered coal and dissolving the resultant barium sulphide in hydrochloric acid by which process barium chloride is obtained. The amount of barium chloride in the solution is determined by its gravity. It is precipitated by a weak solution of sulphuric acid or sodium or magnesium sulphate in the exact quantity to use up all the barium chloride. When sulphuric acid is used, the solution must be mixed cold, but when sulphates are used it is best to conduct the operation hot as a better filtering product is obtained under these conditions.

Blanc fixe is in a great deal finer state of division than the natural sulphate and lacks its grain, a quality which is of great value in plate inks. When used as a base for organic pigments blanc fixe is generally precipitated simultaneously with the coloring matter, using a solution of barium chloride and a soluble sulphate.

PROPERTIES OF BASES

Name	Fineness	Covering Power	Incompatibility	Fastness to Light	Oil Absorption
Natural barytes.	When water floated it is fairly fine.	Fair.	Mixes well with everything.	Fast.	Low.
Paris white.	Somewhat coarse.	Practically none.	With colors that have a trace of acid paris white will cause swelling, e.g. prussian blues.		Fairly low
Blanc fixe.	Impalpable powder.	Fair.	Mixes with everything.		Fairly high.
Aluminum hydrate.	Impalpable.	Practically transparent.	Mixes well with everything.		Very high.
Magnesium carbonate.	Impalpable.	Practically none.	Mixes well with everything.		Very high.

Name	Atmospheric Influences	Shortness	Flow	Smoothness	Abrasive Qualities
Natural barytes.	Is not affected.	Fairly long	Fair.	Grainy.	Somewhat abrasive.
Paris white.		Is short.	None.	Fairly smooth.	Slightly abrasive.
Blanc fixe.		Fairly short	Good.	Very smooth.	Not abrasive.
Aluminum hydrate.		Very long.	Fairly good.	Very smooth.	Not abrasive.
Magnesium carbonate.		Fairly long.	None.	Very smooth.	Not abrasive.

PROPERTIES OF BASES — Continued

Name	Softness	Value as a Base in Ink and Color Making
Natural barytes	Not very soft.	Very valuable as a base for plate inks.
Paris white.	Not very soft.	Valuable principally as a base for plate inks.
Blanc fixe.	Very soft.	Valuable as a base in typographic inks and as a carrier for lakes.
Aluminum hydrate.	Sometimes very soft and sometimes hard and horny. Only the soft hydrate should be used.	Exceptionally valuable as a base for typographic inks and for a carrier in organic lakes. It is also useful in plate printing inks.
Magnesium carbonate.	Very soft.	Is used principally as a base in lithographic and offset inks.

11. ORGANIC LAKES

The organic lakes consist of a dye and a base or carrier. There are three general ways of making these lakes, all other methods, except in some special cases, being modifications of these three.

In the first method the dye solution is mixed and precipitated simultaneously with the base and then fixed.

In the second method the dye is precipitated on the base which has already been made and then fixed.

In the third method, generally used in the case of the insoluble or pulp colors, the dye in the form of a water paste and the base are mechanically mixed together by grinding.

In these different methods the hue of the color can be varied by using different strengths of solutions, different temperatures, different amounts of grinding and in some cases different precipitating or fixing agents. Of course for the various classes of dyes various precipitating agents are employed and it is a little beyond the scope of this work to enter into details regarding them. The general methods of precipitating and preparing these different classes of lakes, however, are as follows:

In the first case a solution of a certain salt of the carrier is made, the dye mixed in and the precipitating agent added with constant stirring. Thus in the case of a dye to be made on an aluminum hydrate and artifical barytes base, a solution of aluminum sulphate is made and a solution of calcined carbonate of soda is run into it, followed by the dye solution. Barium chloride is then added to the whole and the dye is precipitated and fixed. The more stirring the lake is given at this point the firmer it will be fixed. If, however, there is any unprecipitated coloring matter, more barium chloride is added. It is usual to add the precipitating agent in excess, as an excess helps in fixing the dye.

The different hues of these and the second class of lakes also are obtained by a variety of methods jealously guarded by the various manufacturers, the most common of which are different strengths of solution, different temperatures either of the precipitating solution or of the solution after precipitation and by mixing a small amount of another dye in the solution. The first two of these methods also exert an influence on the working qualities of the lake, as the temperature and strength of the solutions have a great deal of influence on the physical characteristics of the base.

In the second case the carrier is made into a thin

paste with water, the dye solution added and, after a thorough mixing, the precipitating solution is run in. Some of these lakes, notably those precipitated with tannin, require the use of a fixative such as tartar emetic.

The third class comprises the insoluble azo colors, which are generally sold in the form of a paste or pulp. These dyes can be ground directly into the carrier on the mill or the dye and carrier can be mixed to a thin consistency and stirred into each other, barium chloride added and the whole boiled. This develops certain changes of hue in some cases and also serves to fix the dyes that are at all soluble in water. Of course where desired the base may be precipitated at the time the dye is poured in. In some cases the dye will not stand boiling with barium chloride so that the precipitation must be made in the cold. The use of calcium chloride and lead acetate as developers instead of barium chloride will give variations in the hues of certain colors.

An advantage that these pulp colors also have is that they can be mixed with the proper base and varnishes and ground directly into printing ink on a roller mill, the water being pressed out in the grinding. In plate printing inks the question of the base for organic lakes is not of great importance but for typographic inks, only those lakes made on a precipitated base, such as aluminum hydrate, blanc fixe, lithopone, etc., should be used. A full discussion of the various bases and their value as ink-making materials will be found in a preceding chapter. In general the precipitating agents employed in manufacturing organic lakes are barium chloride, lead salts, aluminum oxide, tin salts and tannic acid.

In the past very nearly all the organic lakes, that were not to a great extent soluble in oil varnish, were used as printing ink colors and many, even those bleeding in oil and being fugitive to light, which should have barred them entirely, were used. This was especially true of aniline lakes which even now for this reason bear a very bad reputation in regard to fading and bleeding. The wide use of these inferior lakes was in great part due to their superior working qualities and to the fact that they gave bright and fiery colors and hues that could not be duplicated or even approximated in the available mineral pigments.

The use of these unsuitable lakes is now rapidly decreasing, and their place has been taken by the permanent non-bleeding lakes, most of which are made from insoluble azo dyes. As a rule all the lakes made from diazo-dyes, whether the dye is soluble or not, are permanent both to light and atmospheric conditions and do not dissolve in any of the mediums used in printing. These colors are in fact more permanent than any of the mineral colors and are infinitely superior to them in working qualities. They can be obtained moreover in almost every color and hue. The organic lakes can also be made to give a series of very transparent inks, a thing impossible with the mineral colors.

Inks made from the organic lake pigments are absolutely essential to the three-color process, as the colors necessary to produce a properly blended print, from the three plates, in which all the lights and shades have their true color value, can only be made from these pigments; there being no mineral pigments that will give the proper hues of red, yellow and blue.

It is almost impossible, in view of the fact that these lakes are sold under different trade names by the various manufacturers and jobbers, to give any list of colors that would be suitable for use and the only method open to the printing ink chemist in determining the value of a lake is to determine if possible the dye used in making the lake, what the base of the lake is and to practically try out the lake itself for fading, bleeding and working qualities under the scheme already outlined.

An illustration of the difficulties in the way of giving any definite tables or descriptions of the different lakes is shown by the fact that lakes of the same color and hue made from the same dye will be found on the market under perhaps a half a dozen different trade names or numbers. Thus a red called "Scarlet Lake" by one manufacturer may be exactly like the "Brilliant Scarlet" of another manufacturer. On the other hand there is no assurance that the lakes sold by two manufacturers under the same name will be the same hue or even if they have the same color and hue approximately, it does not follow that they have been made from the same dye or in the same manner. One of the lakes may be satisfactory in every respect, while the other may be a very poor color.

SECTION TWO. OILS

Linseed Oil

While there has been a great deal of literature written concerning linseed oil, there is still much work to be done in order to give the technical chemist engaged in manufacturing and using it, a broad enough knowledge for him to do his work to the most economical advantage.

There are very many problems that are constantly arising in practical work with linseed oil, to solve which, the technical chemist must blindly grope and there are many questions purely chemical relating to the testing and identification of this material that are at present mysteries. The old system of constants has proved entirely inadequate to differentiate between oils treated by air blowing or ageing, and adulterated oils. Further there are no definite ways of determining the specific nature of the adulteration even if adulteration can be proved as existing.

The authors have done a great deal of work along these lines but as the field has proven to be so broad they cannot, at this time put forth any theories or deductions; they are, however, convinced that the constitution of linseed oil is far more complex and is affected by many more conditions than the investigators heretofore have imagined.

Linseed oil comes on the market in a number of different grades depending on the quality of the seed, the climatic conditions under which the seed was grown, the amount of admixture of foreign seed, the process of extraction, the length of time it has been stored, the method of treatment if the oil has been refined, and the honesty of those who have handled it either as crushers or jobbers. These conditions cause the physical and chemical constants of the oils produced to vary greatly and also affect their behavior when used for various purposes. The wide variation in chemical and physical constants from the above causes, coupled with the introduction of certain drying oils which have lately come on the market, and new methods of treating non-drying oils, has made it very difficult to determine positively whether an oil has been adulterated or not.

The accepted constants for raw linseed oil are as follows:

Specific Gravity	Not under .931
Iodine Value	Not under 170
Saponification Number	About 190
Flash	Not under 280° C.

As stated before, the constants are influenced by a number of things and the fact that an oil meets the above requirements, or even falls below them to some extent, is no criterion either of its purity or of its adulteration.

The simple treatment of blowing air through the oil at a temperature of about 100° C. changes these constants and at the same time changes the odor and color of the oil completely. Absolutely pure, raw linseed oil blown at 100° C. without any addition of drier or other treatment, has failed to pass the above specification while a fresh pressed oil adulterated with soya bean oil and one adulterated with blown rape seed oil have been found to meet these requirements. The following table shows the result of blowing a pure linseed oil direct from the crusher.

	Before blowing	Blown for 4 hrs.	Blown for 18 hrs.	Blown for 27 hrs.
Specific Gravity	.9355	.9360	940	.946
Iodine Value	184.5	181.4	179.4	161.6
Acid Value	1.0	1.18	1.18	1.18
Saponification Number . .	193.2	193.3	192.9	192.7
Flash.	282	284	282	280

The viscosity was of course increased in proportion.

The figures given below are those of oils adulterated in one case with 10 per cent of blown rape seed oil and in the other case with 20 per cent of soya bean oil:

	Oil as received	With 10% blown rape seed oil	Oil as received	With 20% soya bean oil
Specific Gravity	.9356	.9378	.9325	.931
Iodine Value	178.4	170	181.3	170.6
Acid Value	1.53	2.2	.90	.93
Saponification Number	194.3	194.8	193	192.7
Flash	295	295	296	295

The separation of a flocculent mass on heating an oil to between 270° C. and 300° C., known as breaking, has been found by the authors to be a very erratic thing. Rapidity of heating is a great factor; an oil heated in a test tube where the rise is very rapid will show a break whereas the same oil heated slowly will not. All raw linseed oils, if heated rapidly enough, can be made to show a break even though the oil is bright and clear and has been allowed to settle for six months. The breaking can, however, be eliminated even from a freshly pressed oil by simply warming the oil at 75° C. for a couple of hours. This heating is attended by no change in the physical or chemical constants of the oil.

As a general rule raw linseed oil will separate out a flocculent precipitate on being chilled which will disappear on warming. When oil has been heated to 100° C. and air blown through it, this separation does not take place on chilling.

Raw linseed oil, as it comes from the crusher is not

well adapted to the economical manufacture of printing ink varnishes. All raw linseed oils to be so used, should be subjected to a previous treatment. The most satisfactory treatment has been found to be the blowing of the oil, heated to 100° C., with air. This treatment will produce a fine, clear, odorless oil of high viscosity and gravity and is attended with no loss of weight. The method of blowing is merely to raise the oil to 100° C. by means of a steam jacketed kettle or steam coil, and as this temperature is easily reached, exhaust steam should be used for economical reasons. The air is then blown through the oil in a number of fine jets until the oil has reached the proper viscosity which we have found to be between 1600 and 2000 on an Engler viscosimeter, water being taken as 100.

Oil of this viscosity will take much less time to boil or burn to the different grades of plate and varnish oils and will show far less loss in the burning or boiling. A comparison between the loss of weight and time of burning to medium plate oil between an unblown oil and the same oil blown to a viscosity of 1600 is herewith given:

	Unblown Oil	Blown Oil
Viscosity at start..........	747	1600
Time of burning.. . .	40 minutes	20 minutes
Loss in weight	18%	6%

These oils were burnt to the same viscosity and gravity. The raw oil used was of a light yellow color, with considerable turbidity and a heavy malt odor, while the blown oil was of a rich mahogany color without turbidity or odor. The burnt oil made from the unblown oil was of a greenish color, full of soot, while that made from the blown oil was of a reddish cast, approximately free from carbon. This treatment of the raw oil has been found entirely sufficient and no other treatment is recommended. The color of the oil after blowing we have found to vary greatly, some oils becoming very light in this treatment while others darken considerably, this being no doubt due to the different sources of seeds.

Linseed Plate Oils

In making printing ink vehicles, two methods may be pursued; namely boiling and burning the oil to the proper consistency. If oil is blown long enough it will become the proper consistency but practical trials of plate printing oils made in this way have shown that the oil so made is not satisfactory, having a false body and being too soft. The results obtained by blowing are purely the results of oxidation while the authors have proved that both oxidation and polymerization are necessary to make a satisfactory working plate ink oil or varnish and that this result is only obtained by the use of high temperature. Heavy blown oils are, however, used to a large extent in typographic inks.

Plate oils are generally made in three consistencies, strong, medium and weak, and require less tack and length than typographic varnishes and yield no gloss on drying in the ink. They are therefore made from linseed oil alone without other additions. Since the beginning of the art, the oils for use in plate inks have been burned in iron pots to the desired thickness in the following manner.

The oil is put in an iron pot holding about 5 gallons

and heated to about 160° C. It is then ignited by means of a red hot ladle and allowed to burn with enough stirring to thoroughly mix the oil and to keep the heat from becoming excessive; too much stirring will, however, extinguish the pot. The time required to burn light oil is about 15 to 20 minutes, depending on the initial viscosity of the oil. Medium oil is burnt from 25 to 45 minutes, while the heavy or strong oil is burnt from 2 to 3 hours. On account of the length of time required for the burning of this strength of oil and the consequent heat developed, the burner has to keep the heat down by adding from time to time small amounts of cold oil.

The light oil loses in this process from 5 per cent to 10 per cent of its weight; the medium oil from 8 per cent to 18 per cent and the strong oil from 18 per cent to 25 per cent. As can readily be seen, this great loss is principally due to the use of the oil itself as a fuel, which is a very costly process. It has always been traditional that this burning was necessary to burn the grease out of the oil; and the belief was founded, no doubt, on the fact that a burnt oil will not leave a grease mark on paper. This is, of course, due to the fact that burnt oils have a higher viscosity and gravity and do not spread and penetrate the fibers of the paper.

The burning of the oil is not at all necessary and the polymerization and oxidation required can be effected as well by outside heat, by rapidly raising the temperature of the oil to about 320° C. and holding it there until the oil has reached the proper viscosity and gravity. This process which, instead of utilizing the oil itself as fuel, makes use of the cheaper heating mediums such as gas, coal or electricity and shows a very small loss in the oil. The maximum loss for medium oil being 2 per cent and for strong oil 5 per cent, thus making production of plate oils a great deal cheaper. Moreover the oils made in this way are transparent, of much better color and lack the disagreeable empyreumatic odor of the burnt oils; they are also entirely free from carbon and contain a much lower percentage of free fatty acids.

It has continually been contended that the oils made in this way were not suitable for plate inks but the authors have proved that this is not the case. Plate oils made by this process have been used by us practically and on a large scale and the ink made from them has shown superior working qualities to the inks made from the old process burnt oils.

The apparatus shown in figure 5 was designed by the authors with a view to preventing the oils from accidently boiling over, to keep the loss to a minimum and to carry away the obnoxious fumes arising in the process. Electrical heat was employed on account of its economy, the niceness of control possible with it and because it obviates the dangér of fire.

Soya Bean Oil

The only oil that, at present, in any way compares with linseed oil as a drying oil for use in printing inks, is soya bean oil. This oil dries a little slower than linseed oil and when dried exhibits a film not quite so hard and durable but more elastic than that from linseed oil. The authors found that inks made from soya bean oils and varnishes were almost similar in working qualities and appearance to those made with similar linseed oil vehicles but were just a trifle softer and more greasy. These inks, which were made with drier, dried about as well as linseed oil inks. Soya bean plate oils and typographical varnishes can be used as substitutes for linseed

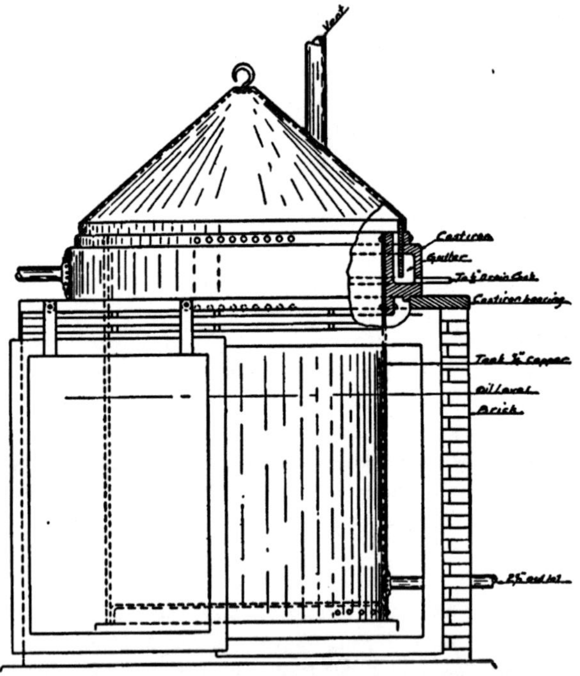

FIGURE 5.—APPARATUS FOR MAKING PRINTING INK VARNISHES.

oil varnishes, except for the highest grade of work. In making printing ink oils and varnishes from soya bean oil, the same processes are used as are employed for making linseed oil plate oils and varnishes.

Soya bean oil is pressed from the soya bean, originally a native of China and Manchuria but now grown throughout the world for fodder. In this country it is widely grown for feeding purposes but as yet no attempts have been made to crush the beans for oil. Its constants are somewhat lower than those of linseed oil and samples from different sources show great variations. Up to the present time there has been very little published about soya bean oil, as its introduction as a competitor to linseed oil has occurred within the last few years.

Samples analyzed by the authors gave the following range of constants:

Specific Gravity	.9240	to	.9270
Acid Value	.70	to	5.2
Iodine Value	124.5	to	148.6
Saponification Number	192.8	to	194

Rosin Oil

Rosin oil and rosin oil varnishes are very important ink-making vehicles, there use, however, being restricted to the typographic inks. They are generally used as substitutes for linseed oil varnishes for the cheaper grades of ink, but are sometines used with linseed varnishes in high grade inks to impart some special property or to meet some special condition of work.

Rosin oil, of itself, is a non-drying oil, drying when used in inks only by absorption into the paper but when boiled with certain driers it forms an oil that dries fairly well. In combination with rosin it makes a good substitute for gum varnishes.

Rosin oil is obtained by subjecting colophony to destructive distillation and is separated by fractionation into four different runs of different consistency and gravity. The first run is the lightest while the fourth run is a heavy, thick, tacky oil of about the same consistency as strong linseed plate oil or No. 4 lithographic varnish.

Paraffine Oil

Paraffine oil is a distillate from the residues of crude petroleum with a specific gravity of from .800 to .820. It is largely used as a vehicle for cheap inks. It has no drying properties whatever except by being absorbed into the paper. Paraffine oil is also largely used in making various special varnishes.

Kerosene Oil

Kerosene oil, in connection with some long varnish, is used as a vehicle for very cheap inks for use on newspapers.

Typographic Varnishes

A. Linseed Oil Varnishes. — The principal typographic varnishes are made from burnt or boiled linseed oil and come on the market in a number of grades of consistency that can be mixed into combinations to suit the character of work for which the ink is to be used, and the physical characteristics of the dry color employed.

The linseed oil varnishes range in consistency from a thin varnish, a little heavier than raw linseed oil, and with about the same drying properties as a painter's boiled oil, to one so heavy that it will not flow till warmed and which dries very rapidly. The heavy lithographic varnishes of linseed oil are very tacky. The varnishes are sold by numbers according to their tack, viscosity, gravity and drying qualities. The thin varnishes range from 0000 to 0 and the heavy ones from 1 to 8. These varnishes are made in the same way as plate oils either by burning or boiling the oil to the proper consistency, tack and drying properties. In some cases for the more rapid drying grades a drier is added such as borate of manganese, red lead or a metallic soap of rosin, linseed oil or tung oil.

A varnish of high gravity but of little viscosity can be made by air-blowing linseed oil at 100° C.; this oil is soft and buttery and has a soft body that makes the ink stand up without being tacky. This together with the fact that it is slow drying makes it an ideal vehicle for offset inks.

It can readily be seen that it is possible by judicious blending of these varnishes to produce inks of the right degree of tack, softness and drying properties for any purpose or process.

B. Gum Varnishes. — Gum varnishes are made from certain gum resins, most commonly damar and kauri. These are melted and mixed with linseed oil heated to a high temperature with or without the addition of driers; this depending on whether a quick drying or moderate drying varnish is desired. These varnishes give gloss to the printed work and a heavy body and great tack to the ink. Inks made with them stand up well on the paper and dry extremely hard. Gum varnishes are always used in combination with other varnishes.

C. Rosin Oil Varnishes. — Rosin oil varnishes are made by heating rosin oil up to 360° C. and adding either borate of manganese, red lead or a metallic soap of rosin. These rosin oil varnishes can be used in place of the linseed oil varnishes for the cheaper grades of ink.

There is also a gum varnish substitute made by dissolving rosin in rosin oil which is known also as rosin oil varnish. In very cheap work a varnish made from paraffine oil and rosin can be used in place of rosin varnish. When properly made this varnish gives good results and is quite cheap.

D. Long Varnishes. — There are certain varnishes made from combinations of asphaltum, rosin, stearine pitch, paraffine oil, wood tar and kerosene oil that are known as long varnishes on account of their length and tack. Different amounts and combinations of the above materials give varnishes of different degrees of tack and length. These varnishes are particularly valuable when it is necessary to use a pigment which in itself is short. Being dark in color they cannot be used except for black or darkly colored inks.

Reducers and Miscellaneous Materials

A. Asphaltum. — The asphaltums are natural hydrocarbons which occur in several forms, the principal differences being in their melting points. Asphaltum is a brilliant, black material with a distinctly brown under hue. As a rule it is hard at ordinary temperatures but there are several varieties that are semi-liquid. Besides being used for varnish making, it is also used to some extent in the manufacture of duo-tone inks, where its brown under hue is taken advantage of.

Asphaltum is also used in preparing lithographic plates

and in making transfer inks. It is readily soluble in all the common solvents and is taken up easily when heated with paraffine oil or kerosene.

B. **Petrolatum.** — Petrolatum is useful for making an ink run smoothly on the press, to correct fast drying in inks that must have a heavy consistency, to take out tack and to shorten inks that are too long.

C. **Soap.** — Soap is also used to make inks work smoothly, to shorten naturally long inks and to give better distribution.

D. **Lanolin.** — Lanolin or purified wool grease will also answer the same purpose as soap but is a little more sticky.

E. **Miscellaneous Materials.** — Beeswax, spermaceti, carnauba wax, paraffine wax, oil of lavender, venice turpentine, balsam fir, balsam of copaiba, gum mastic, stearic acid and olive oil are also used to some extent in making specialities for imparting certain properties to inks but they are mostly used in making different kinds of transfer inks and preparations for lithographic work.

F. **Reducers.** — For reducers to make stiff inks thin, particularly lithographic and offset inks, and to keep inks from drying too fast the following materials are generally used:

The very thin grades of linseed oil varnish.
Rosin oil (particularly to correct fast drying).
Paraffine oil.
Kerosene oil.
Amyl acetate.
Ether.
Oleic acid.

The last three are particularly used in offset work; they should however be used sparingly.

Ether and amyl acetate are used because they are volatile, the theory being that they thin the ink in the fountain but volatilize by the time they reach the work, thus giving an impression that consists mostly of pigment and therefore has more color. That these materials really accomplish this effect is extremely doubtful.

Oleic acid is supposed to assist in preserving the work on the plate. It is used in very small quantities, a few drops to a fountain full of ink.

Driers

The principal driers used in printing inks are the straight metallic driers, such as borate of manganese and red lead, the metallic soap driers which consist of the resinates, linolates and tungates of lead and manganese dissolved in linseed oil, and the japan driers, which consist of a varnish gum and one of the metallic soap driers dissolved in linseed oil and thinned with turpentine.

The first class of driers are used principally in plate inks and they are generally worked into a paste with linseed oil and paris white.

The resinates, linolates and tungates can be made in two ways either by saponifying the rosin or oils, precipitating the acids with sulphuric acid and saturating the free acids with lead or manganese carbonate or by precipitating the soap solution directly with a solution of lead nitrate or manganese sulphate. The soap driers and japan driers are generally used exclusively in typographic work.

There are a number of patented and special driers on the market but it is unnecessary to mention them here as they are merely modifications in some way of the three classes of driers mentioned above and any work claimed for them can be done by the above mentioned driers just as well.

PART THREE

SECTION ONE. THE MANUFACTURE OF PRINTING INK

General Considerations

In the manufacture of printing ink there are four general essentials for producing good inks.

First, the material should be properly proportioned, that is, the pigments should be carefully weighed out so as to make the color desired directly, without additions, and this can generally be done if a formula is worked up in a small way in a laboratory. The vehicles should always be carefully weighed with the error on the side of stiffness rather than softness, as it is easier to make a stiff ink soft than to stiffen up a soft ink, as the latter requires the addition of dry pigment and this is very troublesome, especially if the color is a mixture of two or more pigments. In this case there will always be some difficulty in matching the color. It might be noted that many colors which seem stiff when mixed will be a great deal softer after running through the mills. In these cases experience must be the teacher.

Secondly, the materials must be thoroughly mixed, every particle of pigment be brought into contact with the varnish and the whole mass wet out so that it has a uniform consistency.

Thirdly, the ink should be sufficiently ground to get the necessary homogeneity and smoothness. As will be stated later, some inks from the nature of the pigment used and the kind of work the ink is going to be used on, require more grinding than others. Typographic inks, especially, should be ground to a great degree of fineness and smoothness. In using roller mills, care should always be taken to cut off the ink that runs to the edges and return it to the hopper to be reground.

Fourthly, all inks, after coming from the mills should be blended; that is remixed, so that the color will be uniform. This is particularly necessary in inks that are made from two or more pigments and for pigments that have any tendency to work away from the oil. Corrections in the hue or consistency of the inks should be made at the time of blending. For the purpose of correcting the hue a set of stock inks consisting of the various pigments ground in oil or varnish should be kept on hand. The addition of these will bring the ink to the proper hue and consistency at the same time and obviate the necessity of regrinding the batch, which would be necessary if dry pigments were added. In correcting the hue of inks it should be remembered that not only the top hue but also the under hue should be matched. This is more important in plate printing and halftone work than it is in general typographic work. The main thing to be considered in ordinary typographic work is top hue.

The weighing of the pigments, oils and varnishes can be done on regular platform scales, the pigments being weighed into rectangular sheetiron boxes fitted with handles for lifting. In the case of carbon black, which is so light as to make it difficult to handle, we make the mixing of such a size that the material can be dumped directly into the mixer out of the original bag it was received in without extra weighing.

The oils and varnishes should be stored in tanks fitted with spigots and these materials run directly from the

FIGURE 6.—MIXER FOR MAKING LARGE BATCHES OF INK.

tanks into tared receptacles on the scales. These receptacles should be of such a design as to allow the vehicle to be entirely emptied into the mixer without loss.

For inks to be mixed in a pony mixer, the oils or varnishes should be weighed directly into the mixer cans and the pigments added when the can is on the mixer. Care should be taken that all cans and boxes should be thoroughly cleaned before using. The mixing should be done on one of two regular types of mixers. For large batches and for stock colors, which are regularly made in large amounts, a mixer of the bread or dough mixer type shown in Figure 6 should be used. These mixers should be provided with a close fitting cover and a device for dumping. The authors have found that a mixer of this pattern with gears on both sides, driven by an individual motor with a noiseless chain drive are most satisfactory. For small batches of ink and odd colors not regularly made, where frequent and thorough cleaning is necessary, the type of pony mixer shown in Figure 7 is desirable. This type of mixer is also especially adapted to blending batches of ink which need correction for hue.

FIGURE 7.—PONY MIXER.

The three-roll, water cooled, paint mill is the best mill for grinding inks. This mill, as shown in Figure 8, consists of three hollow rolls of hard steel, two of which, the outside ones, revolve in the same direction but opposite

:om the middle one; these rolls are water cooled to revent the friction heat from affecting the ink and are

FIGURE 8.—MODERN THREE ROLL, WATER COOLED INK MILL, SHOWING AUTOMATIC FEEDING ARRANGEMEMT.

djustable so that different amounts of pressure can be btained. The ink is fed into the back and middle rolls between two movable hoppers and is cut off from the front roll by a steel knife onto a sloping apron. The knife edge should be kept sharp and should always bear

FIGURE 9. — STONE MILL FOR GRINDING INK WHERE A GREAT DEGREE OF FINENESS IS NOT REQUIRED.

true against the roll throughout its whole length. The rolls are geared so that they revolve at different speeds. The most satisfactory way of running the mills is with individual motors and noiseless chain drives.

For grinding large quantities of plate inks to be used

on a power press where a great degree of fineness is not necessary, a stone mill as given in Figure 9 is satisfactory.

Explanation of Terms

The following definitions of terms used in the printing trade to explain certain conditions arising in the work will serve to explain what is meant when these terms are used in the following pages.

Striking Through. — When an ink is very thin or made from a pigment that is soluble in oil, the oil will penetrate through the paper, showing up on the back of the printed sheet either a greasy mark or a discoloration. This is due to the penetration of the oil before it has time to dry, to the fine particles of the pigment being carried into the paper by the oil and to the solubility of the pigment itself in the vehicle. This defect is called striking through.

When work is printed wet, as is often the case in plate printing, this striking through is sometimes due to pigments somewhat soluble in water.

Bleeding is defined in a former chapter as the solubility of a pigment in a vehicle or in water, therefore when striking through is due to either of these causes it can be said to be caused by bleeding.

Offsetting. — When the ink from one sheet comes off on the back of the sheet that is laid on top of it, the ink is said to offset. There is a method of printing called offset printing that will be spoken of later on.

Permanency. — This means the resistance of an ink to light, almospheric and chemical influences.

Picking Up. — When the ink on the plate or form pulls out little fibers from the paper it is said to pick up.

Gathering and Graining. — In plate printing, when an ink collects in little hard or stringy lumps on the roller or ink slab it is said to gather. When the same effect appears on the rollers of a typographic press it is called graining.

Wiping. — The wiping of an ink in plate printing is the way it acts when wiped off the plate by the printer's cloth. After inking in his plate, the printer takes a starched cloth and wipes the superfluous ink off. The ink should come off clean with the cloth but should not come out of the lines.

Polishing. — This is the way the ink leaves the plate when the plate is polished by the printer's hand. After wiping, there is generally a thin film of ink left on the plate, and the printer polishes the plate with his hand to clean this scum off. After polishing, the plate should be perfectly clean without any film of ink on it.

Tints. — In reference to color, the meaning of a tint has been explained in a former chapter as a color lightened with white. In printing terms it is used in a somewhat different meaning, namely to denote an ink printed very faintly and lightly on paper.

Tinting. — When for any reason the ink is taken up on the low parts of a form or when a plate ink has not been polished clean off the plate and this causes a slight discoloration of the whole impression it is said to be tinting. This can be caused by a number of things, chiefly, however, by the lint from the paper filling up the forms, and by the use of a gummy ink with a good deal of light varnish in it, which spreads over the entire plate when the press speeds up.

In the case of plate inks, this is due to a hard wiping or greasy ink.

Rubbing Off. — When a printed sheet is handled after sufficient drying and the ink smears or comes off on the hand, it is said to rub off.

Greasy. — A greasy ink is one that leaves a scum over the plate or form or leaves a greasy margin around the printed work.

Greasiness is due to the use of too thin a vehicle, too thin an ink or to the use of a vehicle that does not dry rapidly enough or to a combination of all three of these defects.

Duo-Tone Inks. — Duo-tone inks are those that show a black over hue and a decided brown under hue, giving the impression that the print has been made with two printings, using different inks.

Printing Inks

Printing inks are divided into two general classes according to the two basic processes of printing. These two basic processes are plate printing, done on intaglio plates, and typographic printing, done either from surface plates such as lithographic plates, or raised plates, or types such as electrotypes and lead types, where the ink-taking surface is raised above the body of the plate. Of course there are many modifications of these two processes in actual use and the ink necessary for these modifications varies with the variations in the process. Therefore under the head of plate printing inks and typographic inks as broad classes will be comprised the various inks suitable for the different modifications of these two general processes.

1. Plate Printing Inks. — Hand-press inks should have a certain amount of length to make them stand up and hold in the lines when wiped. This length is produced by the use of strong oil. Too much length, however, results in a stringy, hard-wiping ink. This can be remedied by reducing the amount of strong oil or in cases where reducing the strong oil will result in a soft ink the base can be increased or a little aluminum hydrate added. Aluminum hydrate will also improve the polishing of an ink when it has a tendency to cause tinting.

Where the ink is too soft and has a tendency to mash out of the lines the strong oil should be increased and the medium oil decreased.

If an ink dries too quickly the drier is cut down. In some cases the addition of medium oil, if this does not make too soft an ink, will also help.

Gathering is generally due to coarse, dry colors and can be corrected by the use of finer raw materials and thorough mixing. In some special cases, the addition of other material will stop gathering. These special cases will be noted under the head of the color when they occur.

Inks that do not fill in well are too stiff and should be softened by the addition of medium oil.

Power-press inks should be made as thin as it is possible to make them without having them mash or pull out of the lines. They only require length enough to make them feed well on the roll and for this reason medium and weak oils are used instead of medium and strong oil, as is the case in hand-press inks. When power-press ink pulls out of the lines when it is wiped, it is generally too short; a little strong oil can be added to correct this. Sometimes this wiping out is due to too dry an ink and it can be remedied by adding medium oil. When an ink does not fill into the plate well and, therefore, prints out light, it can be made to fill in and print better by the addition of medium or weak oil. When an ink feeds badly or lacks distribution a little strong oil can be added to it. If the ink has a tendency to mash it should be stiffened by dropping some of the oil.

For all plate inks the materials having the lowest oil absorptions are the best. When the oil absorption of a color is very great it is of no value as a plate-ink pigment. A pigment of moderately high oil absorption can be mixed with one of lower oil absorption or the oil absorption can be reduced by the addition of more base.

A certain amount of base is necessary in all plate inks except blacks, to give the ink grain and body, to hold it in the lines, to make the impression sharp and clear, to make it wipe well without being too short and to improve the polishing. Without base the ink would fill in badly, pull out of the lines, wipe hard, be sticky and tacky, and tint the plate, making it hard to polish and impossible to get a clear impression.

The tinting of plate inks should be done with zinc white as white lead is heavy, has a tendency to separate on standing and chalks after being printed out. It also works greasy.

In correcting apparent faults of inks, the condition of the starched cloths should be carefully examined as too soft or too stiff a cloth will cause a great deal of difference in the working of an ink, particularly on a power press. The paper also has some influence on the behavior of the ink; a paper that is too wet or not wet uniformly causing a good deal of trouble due to the ink not taking hold of the paper and therefore giving a light impression. Where the water is on the paper in spots or drops the ink at that particular place does not print and the lines will be broken; this is what is known as a waterbreak.

A. Black Plate Inks. — The highest grade of black plate inks are made from pure, acid washed bone black, finely ground, and brought to the proper color by the addition of a high-grade vine black, and enough prussian blue to fully develop the black, but not enough to blue the under hue.

For hand-press ink these materials are mixed to the proper consistency with strong and medium oil; about six times as much medium oil as strong oil being used, and about two pounds of borate of manganese added to every 100 pounds of oil. In mixing the black, the oil and drier should be put into the mixer first and the blacks and blues added a little at a time until it is thoroughly mixed. The mixture should then be ground on a three-roll mill once or twice, according to smoothness.

If the ink shows a tendency to wipe hard the strong oil should be reduced. Too much drying and stiffness can be corrected by adding medium oil. If the ink is too thin and has a tendency to mash add more strong oil. If the addition of medium oil does not correct the drying or makes the ink too thin cut down the drier.

Gathering can be lessened by increasing the medium oil and having the materials ground finer. Coarseness of raw materials is a very prolific cause of gathering. If these two remedies fail to have the desired effect a small amount of magnetic pigment described under special blacks will be found to improve this condition.

Sometimes blacks have a tendency to puff up on the mill and to become spongy; this is due to occluded air in the black and it generally takes place in air-separated blacks. There is nothing that can be done to remedy this condition except to take care that the air is entirely gotten out of the black in packing.

All the materials that go to make up blacks should be low in oil absorption. The black pigments used should not require more than 65 parts of medium oil to 100 parts of color.

In making black ink for use on power presses, the strong oil should be left out entirely and its place taken by about half as much weak oil as there is medium oil. The borate of manganese should also be increased about ½ pound for every 100 pounds of oil. The above formula will give a very satisfactory black plate ink, the cost and grade depending on the kind and value of the bone and vine blacks used.

Plate inks made from carbon or lampblack will not work satisfactorily as they lack grain and body and have a very high oil absorption.

B. Green Plate Inks. — Fine green plate inks, made from chrome green should be lightened with chrome yellow lemon. Chrome green being very dark in itself, it is never necessary to shade it. For cheaper work the green can be lightened with barytes but this produces a pale tint with blue hue that is liable to chalk. When yellow is used to lighten green it gives a green of light top hue and slightly yellow under hue. If a very light green tint is required zinc white can be used with yellow as a diluting agent. In order to make a green plate ink, made with chrome green and chrome yellow lemon work satisfactorily, it is necessary to use a base; this should be a water-floated barytes with about 10 per cent of paris white in it. A good chrome green ink will stand about 50 per cent of this base. The base makes the ink wipe and polish well and keeps it from rubbing off. Chrome green plate inks dry well naturally and in most cases it is unnecessary to add drier.

The green used for hand presses should have about 1½ pounds of strong oil to every 100 pounds of medium oil to prevent it from flowing out on the stone. While green plate inks are always used softer than black inks, care should be taken not to get the consistency too thin as this will cause mashing. If the green wipes badly the strong oil should be decreased or the base increased a little. A slight increase of base also corrects tinting, a fault inherent in all greens. When the green ink gathers the medium oil is increased.

In mixing greens, the green, yellow and part of the medium oil should be mixed and ground first, being put through the mill once. Then the base, the strong oil and the rest of the medium oil is added to the mixture and the whole mixed thoroughly and reground. For power-press inks the strong oil is left out and enough weak oil to bring the ink to the proper consistency is used.

Up to the present time there are no very satisfactory green lakes on the market for use in plate inks.

Olive green can be made by mixing prussian blue with the medium hue of chrome yellow. This gives a bright olive; it can be shaded with a little vine black or lightened with a little chrome yellow lemon. Its base should be barytes. Olive inks should be ground about three times to prevent the separation of the blue on standing.

C. Yellow Plate Inks. — The chrome yellows give very satisfactory working plate inks but when a fine permanent ink is desired one of the permanent yellow lakes will be found not only to work better but also to be more permanent and brighter than the chrome colors. The chrome yellows will carry about 50 per cent of base while the yellow lakes will carry slightly more, being stronger and brighter. The question of price and the class of work figures largely in the choice between the chrome yellows and the yellow lakes.

D. Blue Plate Inks. — The blue mineral pigments all make good working inks of fairly low oil absorption, except ultramarine blue which works up into a tacky, hard-working ink. There are, however, very permanent blue lakes without these faults that will match ultramarine blue in color. There are a number of blue lakes of all hues on the market that surpass in working quality and permanency all of the mineral blues, but they are quite expensive as compared with the mineral blues. A small amount of drier should be added to all the blues.

E. Red Plate Inks. — Among the red inks there are no satisfactory ones made from mineral pigments. Vermillion and orange mineral, the latter not being really a red but an orange, make very poor working inks. Vermillion is a very heavy pigment and settles out from the ink and on this account causes the ink to gather a great deal and to fill in poorly, defects that it is impossible to remedy. Where vermillion must be used it should always be used with some similar hued pigment. Vermillion inks always require driers.

Orange mineral ink has the same defects as the vermillion ink with the added disadvantage that this pigment exerts a great drying action on the oil, causing the ink to harden up rapidly. It also has a tendency to form a soap with the oil, which results in a livery mass that cannot be worked. Of the two grades of orange mineral, French and German, the French orange mineral is the better.

Since there are a number of absolutely light and atmospherically fast aniline lakes of all hues of red, from a brilliant scarlet to an orange of red hue, which possess perfect working qualities, it is an easy matter to match the color of any red. The red lakes will carry a great deal of base when made into ink. The most common fault of the red lakes is shortness and care should be taken to avoid those colors that have a high oil absorption or show shortness. The addition of a little more base will make an ink of this character work better, if it is necessary to use a pigment that is short. All red lake inks should be made up with a drier.

F. Degraded Colors. — In making degraded colors the pigments to be mixed must be selected with a view to avoiding colors that will act on each other chemically. They should also be selected with reference to their specific gravity and oil absorption. Colors that are close to each other in specific gravity should always be selected, as this will prevent the separation of the two colors on standing. When pigments of widely different specific gravity are used, this separation will take place no matter how intimately the ink is mixed or how often it is ground. Two colors of relatively high oil absorption should not be used together but an attempt made to secure an average oil absorption where it is necessary to use a pigment that takes a great deal of oil by combining it with one that takes very little.

In mixing inks of degraded colors a set of stock colors, consisting of the different materials needed in the mixtures and ground in medium oil, should be kept on hand and the final adjustment of top hue and under hue made by adding from these stock inks. Care should be taken in making this final adjustment that the under hue is not changed too much in attempting to bring up the top hue and vice versa.

In plate printing, both the under hue and top hue must be taken into consideration as they are both brought out in this work and it is necessary to match a sample in both these particulars. The fewer pigments used to obtain a color effect the better the ink will be and the easier it will be to match at some future time.

In matching or comparing samples printed at two different times, allowance must be made for the time the original sheet has been printed, as ageing and drying out will affect all inks to some extent, especially the under hue. The condition of the plate also has an effect on the color and work done when a plate is new will always look much brighter and stronger than that done on a worn one.

2. Typographic Inks. — Typographic inks are divided into a great number of special classes, according to the effects to be produced, different processes for producing these effects and different styles of presses. Even inks to be used to accomplish the same general effect have often to be made slightly different from each other on account of variations in the methods of use or in the kind and quality of paper employed.

Basicly typographic inks consist of a pigment and a vehicle, all other ingredients being added to produce some effect or remedy some defect due to the conditions of the process to be employed or the style of press; as for instance to increase or decrease the drying, to give a gloss; to increase or decrease the body; to give length or shortness. In all inks, especially typographic inks, complicated formulas should be avoided and the materials and effects desired so studied that a certain result is reached with the simplest possible mixture of color and vehicle.

A. **Pigments.** — The best pigments for use in typographic work are the light and atmospherically fast organic lakes, made on aluminum hydrate and blanc fixe; and as these can be obtained in a number of different hues and colors it is seldom that two of them have to be mixed to produce a certain color. This is of great importance, as frequently the mixture of two or more colors of different gravity, oil absorption or working qualities will result in a separation taking place in the fountain, which will cause trouble in the feeding and distribution of the ink. This is particularly true of the mineral pigments which should never be used in typographic inks except for the cheaper grades. The mineral colors are as a rule satisfactory for plate printing inks as there is not the demand in that sort of printing for the working qualities, the very brilliant colors and the variety of hues that are called for in lithographic and typographic inks.

In typographic printing there is a demand for brilliant colors of almost every hue, many of which must be transparent and the only way that inks meeting these conditions can be produced is by the use of organic lakes, principally those made from aniline dyes. In cases where the ink manufacturer makes his own lake colors, the best procedure and one that gives the most brilliancy is to grind the lake directly into varnish as it comes from the filter press without any drying, as drying always has a tendency to dull the color and in some cases even changes the hue of the lake. Where a lake color is to be dried before use, the latest factory practice is vacuum drying which does not affect the color nearly as much as drying by heat alone and which does not require nearly as much care and attention to produce a uniform product. In grinding moist lakes into ink, the lake should first be ground in a small amount of heavy varnish until all the water has separated out and then ground with the other varnishes to the consistency desired.

For producing tints in typographic inks, zinc white should be used while for making transparent ink, aluminum hydrate is the proper material. In making a tint in typographic work a very thin ink with just enough tack to make it take on the rollers and distribute, is most satisfactory and in many cases tint inks are dyes dissolved in the vehicle.

B. Vehicles. — The use of the different vehicles is dependent on the class and cost of the ink desired. In general, for all high-grade inks, linseed oil varnishes of different consistencies are used as the basic vehicle. In cheaper inks rosin oil is used, while in very common inks such as newspaper and handbill inks, petroleum oils into which some paraffine wax has been ground, are used.

There are a number of grades of linseed oil varnishes

of different consistencies as noted in the preceding chapter on oils and varnishes. These run in consistency from an oil slightly heavier than raw linseed oil up to a stiff almost glue-like mass. Some of these varnishes can be used alone but a typographic ink usually contains a mixture of two or more of these grades of linseed oil varnish and some other varnish such as a gum varnish, japan drier, asphaltum or some special product, added to give a property not possessed by linseed oil alone. The simpler, however, this mixture of varnishes is made the better will be the resulting ink and the more pronounced will be the effect desired.

Rosin oil comes in several different consistencies which are used in the same way as linseed oil varnishes, but for a different gràde of work. Rosin oil is also used to correct a fast-drying ink made from linseed oil varnishes. Rosin oil, even in the cheaper grades of ink, is generally used with some linseed oil varnish, as when used alone it does not dry of itself except by absorption into the paper and dries only to a certain limited extent when used with metallic driers. Thus, when made without linseed oil or a metallic drier rosin-oil inks rub off somewhat, as the vehicle is absorbed and the pigment is left on the surface without much of a binder. This is especially true when the printing is done on an absorbent paper. The use of rosin-oil inks without driers is nevertheless quite common.

Linseed oil varnishes dry partially by oxidation, that is besides being absorbed to a certain extent by the paper, the varnish itself dries in a film around the pigment so that it will not rub off.

Gum varnishes are used to give tack, body and gloss.

They also tend to make the ink stand out on the paper somewhat and when used with linseed oil varnishes they produce an increased drying effect. The gum varnishes are used with oil varnishes either alone or in combination, depending on the effect to be produced and the pigment used in the ink. No general rule for the use of the different kinds of gum varnishes can be laid down as the effects produced by them vary somewhat with the process, paper and style of press used. Experience in producing certain effects with certain combinations is the best guide for their use. In using gum varnishes, however, as small an amount as possible to produce the effect desired should be used. Rosin varnishes and rosin-oil varnishes are used to give the same general effects as gum varnishes for cheaper inks; and are also used to correct the fault of drying in inks which must have a rather heavy body.

C. **Driers.** — Driers are used when it is necessary to increase the drying of an ink without changing its consistency, length or tack and to make extremely quick drying inks. An ink that dries slowly and offsets can be made to dry more rapidly by the addition of a gum varnish and to dry quickly by the addition of a drier. A gum varnish exerts only a moderate drying action while a drier will make an ink dry rapidly. The addition of a gum varnish is apt to change the consistency or the properties of an ink so that if a moderate drying action is desired and the addition of a gum varnish will make the ink too stiff or tacky a drier can be added and the drying regulated either with rosin varnish or rosin oil. As a general rule the addition of the so-called reducers and specialties should be avoided; however, the use of soap and petrolatum is of great assistance in making inks soft or to make them work smoothly and cover well when made from an inferior color.

The class of varnishes called long varnishes are used to give length and flowing qualities to inks. They are generally made on an asphaltum or wood tar base. There are a number of ingredients that go into long varnishes which are used in cases where the ink demands greater or less length, tack and body. The demand for these properties determines to a great extent what the composition of the varnish will be.

It must be remembered in making all inks that frequently, with the same raw materials, a formula successful in one case will give slightly different results in another, especially under different weather conditions. This matter of weather condition is one that in the working of inks has not been given any very serious consideration, but it is a thing that plays a very important part in the proper manipulation of many typographic inks.

D. **Offset Inks.** — For offset work the inks should be short and have fairly little tack; the best inks are rather stiff, that is, they carry a great deal more pigment in proportion to the vehicle than the ordinary typographic ink, and should be somewhat buttery in consistency although they should not be too greasy as the grease tends to form a tint on the plate. They should also be quite slow drying, as an ink that dries even moderately fast will make frequent washing of the plate necessary and every time it is washed the life of the plate is shortened. An ink that dries fast or that is sticky also piles up on the plate and causes the lines to spread, which is also a serious defect. The ideal ink for the offset press is one that has only a moderate drying action, that has only enough tack in it to feed well and distribute properly, and has such a body that the greater part of the ink that goes on the paper consists of pigment, having, however, enough vehicle to hold the color without rubbing. As the amount of ink that goes on the plate is quite small, it can easily be seen that to get a properly colored print a great deal of pigment must be carried in a relatively thin layer of ink.

If the ink is too stiff it can be reduced with a weak lithographic varnish, boiled linseed oil without driers or with thin blown oil. Sometimes it is necessary to use petrolatum or paraffine oil, but only small amounts of either of these should be used as they have a tendency to spread the work on the plate and to make the printing strike through. In cases where the ink is too stiff and tacky, petrolatum, mutton tallow or lanolin can be used in small quantities with good effect. Ether and banana oil are both recommended to make the ink distribute well and give greater color, but our experience has been that an ordinarily soft ink made from proper pigments, reduced when necessary with petrolatum or paraffine oil, will accomplish more than these volatile solvents which for the most part are dissipated before the ink leaves the fountain and which are not healthy additions to the heavy air of the average unventilated press room.

We have found that the best basic vehicle for offset inks is linseed oil, thickened to about the consistency of No. 1 plate oil by blowing air through it at a temperature of about 100° C.

A first-class grade of carbon black should always be used for blacks and lakes precipitated on very soft-grained aluminum hydrate should be used for colors. An ink that separates out color or pigment to the very slightest extent should be avoided as the slightest piling on the plate makes a very bad looking job. The latitude allowed on flat-bed or rotary cylinder presses using electrotypes cannot be allowed in an ink for offset work.

For tints in offset work and in fact for any colored work except black, a base consisting of equal parts of magnesium carbonate ground in a thin varnish to a stiff paste and a mixture of zinc white and aluminum hydrate also ground in varnish will be found not only a good reducer but also to give the necessary body and working qualities to the ink.

SECTION TWO. DEFECTS OF INKS AND THEIR REMEDIES

The usual difficulties met with in using typographic inks and their remedies are as follows:

Working away from the ink rollers.
Lack of distribution.
Drying on the rollers.
Offsetting.
Flooding the type.
Picking up.
Filling the forms.
Tinting the forms.
Rubbing off after drying.
Graining on the roller.
Drying too fast.
Not drying fast enough.

Working away from the Ink Rollers in the Fountain. — Frequently an ink will work away from the feed roller in the fountain and the result will be that the plate or forms do not get the proper amount of ink or the ink that is fed, is not evenly distributed over the form or plate and a poor print results. This is due primarily to a short ink and this condition may be caused either by the color mixing short or the use of too great an amount of short varnish. The remedy for both these conditions is to add a certain amount of varnish that has length. In cases, however, where the pigment is in itself inherently short in all varnishes, it is best if possible to use some substitute of the same color and hue, that works properly. Where the presence of a varnish contributes to this shortness it is best, besides adding a varnish that will give length, to drop as much as possible of the varnishes that are causing or contributing to the shortness of the ink, where this can be done without affecting the results desired in the finished work. This lack of distribution or shortness is sometimes developed in an ink on standing but can be destroyed by restirring the ink. This is probably due to some tension in the particles of pigment and is most frequently encountered in certain grades of carbon black.

Lack of Distribution. — It is sometimes the case that an ink will work well and feed from the fountain nicely but will not distribute on the ink bed or rollers. Besides being due to shortness this is sometimes due to a soft ink and in these cases the tack should be increased to such an extent that the ink will take hold of the rollers easily and spread over all their surface. Sometimes, however, this lack of distribution is caused by too rapid drying, the remedies for which are taken up in the next paragraph.

Drying on the Rollers. — Drying on the rollers is due most often to an ink that dries too rapidly but sometimes this is caused by an ink that is too tacky and for this reason picks up and holds particles of lint or dust, and it is not infrequently caused by conditions of the atmosphere; so that an ink that works perfectly under the ordinary conditions will dry rapidly in a spell of very dry weather. It is best to decide first whether the drying is the result of tackiness and if this is the case the rollers and the ink on them will contain some fine particles of lint and dust and not make a clean dried film but a gummy mass, and the sheets are apt to show where the paper has been picked up. If the trouble is from this source the remedy lies in making the ink softer. If, however, the drying is due to a quick drying ink, that is, if under any weather conditions the ink dries in a clear film, the drier should be cut down or if there is no drier in the ink and the drying is due to the influence of the pigment used a small amount of non-drying oil should be added. If the ink, which has been running well under ordinary weather conditions suddenly begins to dry too fast this condition can be remedied by adding a little non-drying varnish to the ink in the fountain to meet the change in the weather; inks that dry for this cause should always be doctored in the fountain as this condition may change at any time.

Offsetting. — Offsetting may be due to a number of things that must be taken up in order. It may be that the ink is not getting its initial set fast enough, in which case the amount of drier should be increased. It may be also due to lack of absorption of the paper, which may have two causes: either the ink may be too stiff, in which case it should be made thinner or it may be due to a poorly made paper for which trouble there is no way of improving the ink. Offsetting is also caused by the printer allowing too much ink to go on the form and this can be stopped by cutting down the feed of the fountain. A sticky ink is also a cause of offsetting and this can be stopped by adding petrolatum to the ink and where gum varnishes are being used, by cutting down the proportion of these varnishes.

Sometimes, especially in the case of high-speed web presses and offset presses a little paraffine or beeswax will remedy this defect.

Flooding the Type. — Flooding the type or form is caused by the printer opening his fountain too wide, the use of too thin an ink or in the case of lithographic or offset work the use of too little water or the absence of an etching material in the water fountain. Where the work in lithographic or offset printing does not come out clear, it is well to add some etching solution to the water fountain.

Picking Up. — Picking up is caused either by too stiff an ink, too tacky an ink or an ink that dries too fast and, therefore, develops tackiness on the rollers, and also by poor paper. When the ink is stiff it should be thinned by the use of a weak varnish; when it is too tacky petrolatum or a soft varnish should be added and if the ink dries too fast the decreasing of the drier or the addition of a non-drying varnish should help. Even when the defect is caused by poor paper the trouble can be remedied to some extent by the use of the different materials described above. Of course this must be the subject of experiment.

Filling the Forms. — Filling up of the forms or type may be due either to too rapid drying or the use of a pigment that does not distribute well in the varnish. This generally occurs when the ink is not ground well enough to make it smooth, when the pigment itself is coarse or when a pigment of heavy specific gravity is used. Filling up is also caused by the accidental admixture of some foreign matter such as lint from the paper, dirt and dust from the air and the like.

Tinting the Form. — When an ink has a pigment of high gravity and a great deal of weak varnish in it or when the ink is too thin bodied, the varnish will sometimes separate from the pigment, particularly when the rollers become heated after a long, continuous run and this varnish, faintly colored, will fill in between the lines and leave a faint tint on the paper. To correct this fault, which is most prevalent on high-speed presses, it is necessary to give the ink more body.

Graining out on the Rollers. — Tinting is generally associated either with the filling up of the form or the depositing of the pigment in a grainy condition on the vibrating rollers. This graining out on the rollers is caused by the separation of the pigment, due to the character of the pigment as explained under "filling the form" or to the same cause as that of tinting, namely the use of an excess of a thin-bodied varnish. The most satisfactory remedy for all three of the above conditions is to use a better working pigment or a heavier bodied varnish of little tack.

Rubbing Off. — There are two varieties of rubbing off,

one where the ink rubs off wet, and this can be remedied by adding more drier, and another where the ink comes off as a dust, due to the absorption of the vehicle into the paper, leaving the pigment on top in a powdery form.

This can best be remedied by the use of a small amount of vehicle that dries by oxidation or the addition of soap, or some other material that will tend to keep the vehicle from penetrating into the paper to such an extent that it leaves the pigment no binder.

This kind of rubbing off is also caused by the particles of coating, sizing or loading of the paper flaking off. Some pigments have the tendency to dust inherent in themselves. In these cases the pigment or paper should be changed.

INDEX

INDEX

L

M

O

P

R

S

T

Printed in the United States
217254BV00004B/7/A

9 781103 807680